Highway Accidents

Highway Accidents

Investigation, Reconstruction and Causation

Bertan W. Morrow, Ph.D.

Copyright © 2016 by Bertan W. Morrow, Ph.D.

All rights reserved. No part of this publication may be reproduced, stored in a retrieval system, or transmitted, in any form or by any means, except as may be expressly permitted by the 1976 Copyright Act or by Bertan W. Morrow, Ph.D. in writing.

ISBN: 978-0-692-60125-9

Whilst every effort has been made to ensure that the information contained within this book is correct at the time of going to press, the author and publisher can take no responsibility for the errors or omissions contained within.

Printed by Createspace
Typesetting and cover design by *www.wordzworth.com*

Acknowledgements

I express my gratitude and appreciation to those who were so helpful in the development of this book. The drawings used throughout were developed with the tutelage and guidance of Roxana Macias. The many hours of editing by Lynn and Toby Berk produced a more orderly and readable text. Finally, without Betty's encouragement, advice, and work this book would never have been written.

Contents

List of Figures   v

List of Tables   ix

Symbols and Abbreviations   xi

Foreword   xiii

1. **Introduction**   1

    Purpose

    Qualifications

2. **Evidence**   9

    Witness Statements

    Collection of Physical Evidence

    Interpretation of Physical Evidence

    Coefficient Of Friction

    Reaction Time

3. **Vehicle Dynamics**   43

    Steering

    Acceleration

    Deceleration

    Slow Yaw

    Post-Impact Spinout

    Rollover

4. Impact Dynamics — 85

Fixed Barrier Impact

Vehicle To Vehicle Impacts

Occupant Dynamics

5. Analysis Procedures — 109

Post-Impact

Impact

Pre-Crash

Collision Configurations

Critical Speed

Vault

Summary

6. Examples — 153

Car vs. Car Intersection Collision – Quantitative Analysis

Truck vs. Suv Intersection Collision

Car vs. Car Intersection Collision – Qualitative Analysis

Roadside Collisions

7. Reconstruction Technology — 191

Computer-Assisted Analysis

Event Data Recorder

8. Causation Summary — 205

Human Factors

Vehicular Factors

Environmental Factors

Causal Interactions

Injury Causation

9. Countermeasures — 223

Drivers

Vehicles

Streets and Highways

Injury Mitigation

A.1. Physics Review — 237

Kinematics

Dynamics

Rigid Bodies

Solid Mechanics

Spring Forces

List of Figures

Figure 2.1.	Change in Acceleration (Jerk)	12
Figure 2.2.	Adjustable Vehicle Jig	18
Figure 2.3.	Curve Radius	18
Figure 2.4.	Slope	19
Figure 2.5.	Slow Yaw – Tractor Trailer	21
Figure 2.6.	Damage Profile	22
Figure 2.7.	Skidmarks	24
Figure 2.8.	Dual Wheel Rear Skids	24
Figure 2.9.	Yaw Mark	27
Figure 2.10.	Redirected Front Tire Skid	28
Figure 2.11.	Impact Tire Marks, Gouges and Fluid Trail	30
Figure 2.12.	Fluid Trail	30
Figure 2.13.	Debris Field	31
Figure 2.14.	Frontal Damage	32
Figure 2.15.	Side Impact with Tree	33
Figure 2.16.	Sidesway	34
Figure 3.1.	Locating Center of Gravity	52
Figure 3.2.	Partial Braking	57
Figure 3.3.	Trailer Without Brakes	61
Figure 3.4.	Tire Forces – Braking and Steering	64

Figure 3.5. Partial Braking – Stability	65
Figure 3.6. Slow Yaw	66
Figure 3.7. Slow Yaw – Passenger Car	67
Figure 3.8. Slow Yaw – Tractor Trailer	67
Figure 3.9. Spinout	69
Figure 3.10. Impending Tipping	76
Figure 3.11. Tripping	79
Figure 3.12. Geometry of Tripping	79
Figure 3.13. Rollover – First Contact	82
Figure 4.1. Speed vs. Crush	87
Figure 4.2. Test Profile vs. Accident Profile	91
Figure 4.3. Test Profile vs. Accident Profile	92
Figure 4.4. Narrow Impact	93
Figure 4.5. Angle Impact	94
Figure 4.6. Angle Impact	97
Figure 4.7. Axle Impact	100
Figure 4.8. Graphical Solution	104
Figure 4.9. Principal Direction of Force	106
Figure 5.1. Speed Correlation	113
Figure 5.2. Acceleration	120
Figure 5.3. Head-On Collision	127
Figure 5.4. Underride Collision	132

Figure 5.5.	Redirection Gouges	133
Figure 5.6.	Override Collision	133
Figure 5.7.	Multiple Vehicle Collisions	136
Figure 5.8.	Car – Motorcycle Collision	137
Figure 5.9.	Pedestrian – Car Impact	139
Figure 5.10.	Bicycle – Bridge Encounter	142
Figure 5.11.	Tanker Rollover	143
Figure 5.12.	Vault	147
Figure 5.13.	Vault	148
Figure 6.1.	Car vs. Car Intersection Collision – Quantitative Analysis	154
Figure 6.2	Truck vs. SUV Intersection Collision	160
Figure 6.3.	Car vs. Car Intersection Collision – Qualitative Analysis	168
Figure 6.4.	V1 at Rest	169
Figure 6.5.	V2 at Rest	169
Figure 6.6.	V1 Fluid Trail	170
Figure 6.7.	Impact Configuration	171
Figure 6.8.	Rollover on Slope	174
Figure 6.9.	Failed Median Barrier	177
Figure 6.10.	Rigid Barrier Impact	178
Figure 6.11.	Frontal Impact	179
Figure 6.12.	Energy Absorbing Device	181
Figure 6.13.	Breakaway Pole	184

Figure 6.14. Pavement – Shoulder Drop Off 188

Figure 7.1. Acceleration Pulse 196

Figure 7.2. Speed vs. Time 197

Figure 7.3. Intersection Collision 198

Figure 7.4. Post Impact Speed 200

Figure 7.5. Spinout Acceleration Pulse 201

Figure 7.6. Spinout Speed Loss 201

Figure 8.1. Accident Causation Model 206

Figure 8.2. Misplaced Sign (On Ground) 220

Figure 8.3. Conflicting Markings 220

Figure 9.1. Mirrors – Exterior View 227

Figure 9.2. Mirrors – Interior View 228

Figure A.1. Projectile Motion 240

Figure A.2. Circular Motion 242

Figure A.3. Friction Force 249

Figure A.4. Moment of a Force 250

Figure A.5. Couple 250

Figure A.6. Load vs. Deformation 254

Figure A.7. Force vs. Crush 255

List of Tables

Table 2.1. Coefficient of Friction for Locked Wheel Skids   36

Table 3.1. Vehicle Acceleration Values in Ft/Sec2   47

Table 3.2. Drag Factors for Passenger Cars   48

Table A.1. Quantities Describing Translation and Rotation   252

Symbols and Abbreviations

Symbols

a = acceleration
c = crush
e = super elevation or coefficient of restitution
f = coefficient of friction, friction coefficient, drag factor
g = grade or acceleration of gravity
h = height
j = jerk
k = spring constant
l = length
m = mass or middle ordinate
n = number
p = percent
s = distance along curve
t = time
x, y, z = coordinates
F = force
I = moment of inertia or impulse
L = length
M = momentum
N = normal force
R = radius
S = speed
T = torque
V = velocity
$V1$ = Vehicle #1
$V2$ = Vehicle #2
$V1-1$ = Driver of $V1$
W = weight
$W1$ = Witness #1
X, Y, Z = coordinates

Abbreviations

in = inches
ft = feet
sec = second
ms = millisecond
min = minute
mph = miles per hour
lb = pound
KE = kinetic energy
PE = potential energy
CG = center of gravity
PDOF = principle direction of force

Foreword

Bertan Morrow holds a Bachelor of Civil Engineering, a Master of Science in Civil Engineering, and a Ph.D. in Mechanical Engineering. He taught undergraduate and graduate courses in Civil Engineering for over 20 years at the University of Miami. While on the faculty he spent approximately 10 years conducting research for the National Highway Traffic Safety Administration (NHSTA). The initial purpose of this research was to develop field investigation procedures, as well as methodology for the reconstruction of traffic accident events. This was followed by the identification of accident causation factors, injury causation mechanisms, and the promulgation of countermeasures (human, vehicular and highway) to reduce accident frequency and injury severity. This multidisciplinary research entailed the on-scene investigation of several hundred serious highway crashes where the integrity of the scene was being preserved by police agencies.

A subsequent four-year study, involving a statistical sampling of all the accidents (including all fatalities) in a specific geographical location, yielded a subset of several thousand accidents. The study of these accidents involved next-day site visits, timely vehicle inspections and the collection of injury data. Several computer reconstruction and simulation programs were tested and evaluated. This research effort was coordinated with similar federally funded studies conducted at other locations around the country.

Additionally, Morrow taught courses and conducted research in highway design and safety and developed the initial edition of the *Manual of Uniform Minimum Standards for Design, Construction and Maintenance of Streets and Highways* for the state of Florida.

Other activities include teaching accident reconstruction classes for various Federal, State and local agencies. As a consultant, Morrow has been involved in excess of 3,000 accident cases regarding civil and criminal matters.

Introduction

Our nation's highways have evolved into a vast, complex system of roadways that are inadequate, deteriorated and obsolete. The vehicles involved – automobiles, busses, trucks, motorcycles and bicycles –are somewhat incompatible, in various states of disrepair, and are typically operated by drivers with minimal training and regulation. The entire system is supervised and regulated by thousands of different federal, state and local agencies with disparate, often conflicting, responsibilities and objectives.

This highway system, like all systems, is subject to failures, ranging from woefully inadequate capacity in all our major cities to the deterioration and collapse of the physical infrastructure throughout the nation. Vehicular collisions, however, represent the most serious system failure. In addition to billions of dollars in damage, the deaths and injuries resulting from these crashes cause untold pain and suffering. Nearly everyone has been affected by death or injury to some family member or friend. During the 40 plus years I have been working in this field over 1,500,000 people have been killed and millions more injured on U.S. streets and highways. This is more than twice the total number of Americans killed in all the wars in the history of our country. More are killed in highway crashes each *month* then died in New York on September 11, 2001. Looking at it from another perspective, more people are killed by vehicles than by guns.

In order to reduce the consequences of these system failures we must develop and implement effective countermeasures, not only to limit the frequency of accident events but also to reduce their severity. These corrective measures will need to be directed toward improved driver regulation, vehicle design and maintenance, and the design, construction, maintenance and operation of the highway network. Implementation of some relatively simple, effective countermeasures could most probably reduce deaths and serious injuries by one half and the cost would likely be far less than the hundreds of billions of dollars our nation has spent in response to 9/11.

In order to develop appropriate countermeasures we need to understand the factors contributing to the causation of crashes as well as those factors affecting injury severity. We need to identify how and why the system failed. Research efforts utilizing appropriate investigation and reconstruction procedures are required to provide the basis for the determination of these causal factors. This reconstruction effort depends upon the collection and proper interpretation of the available evidence.

The entire process can be summarized as follows:

- Collection of evidence
- Interpretation of evidence
- Qualitative reconstruction
- Quantitative reconstruction
- Causal analysis
- Promulgation of countermeasures
- Implementation of countermeasures

This comprehensive research effort requires the participation and cooperation of people with a variety of backgrounds and expertise, including (but not limited to) police, investigators, analysts, attorneys, judges, social scientists, engineers and physicians. It also requires the participation and cooperation of public safety agencies, auto manufacturers, highway agencies and the medical profession.

In addition to research efforts, criminal and civil court actions provide important incentives for countermeasures on a case-by-case basis. These actions can provide an important means of regulation by penalizing poor driver behavior, defective vehicle design, substandard highway design, unacceptable construction practices and inadequate maintenance and operation activities. This court action must be preceded by the same process required for research efforts in arriving at the cause, or causes, of the specific accident event and resulting injuries.

Purpose

The purpose of this book is to:

- Assist the accident investigator in the utilization of priorities and procedures that identify and properly document physical evidence through improving the ability to interpret physical evidence and understand the type of information that is critical for the reconstruction and causal analysis.
- Provide guidance to those qualified individuals entering the reconstruction field regarding field investigation, interpretation of physical evidence, application of physics principles, and the use and limitations of various reconstruction techniques.
- Share with qualified and experienced individuals working in this field my thoughts regarding reconstruction priorities and procedures including suggestions for the analysis of some accident situations they may not have yet encountered.
- Present an orderly process for identifying the cause or causes of an accident, including the interactions of human, vehicular and highway deficiencies.
- Provide a non-medical look at occupant kinematics and the associated physical evidence required to evaluate injury causation mechanisms.
- Propose some countermeasures that would be effective in reducing the frequency and severity of highway crashes.

The achievement of these objectives should be beneficial to police officers, accident investigators, reconstruction and causation analysts, vehicle and highway engineers and transportation agencies. Attorneys and judges might also have some interest in the proposed countermeasures.

This endeavor is not intended to be a comprehensive in-depth textbook, a rigorous scientific exercise, a compilation of accident reconstruction data or a summary of relevant research. It is my effort to share what I have learned through my education and experience and collaboration with others in the field. My research, teaching and consulting in field investigation, reconstruction, causal analysis and promulgation of countermeasures provide the basis for the content of this book. I will be addressing what I feel are the important priorities and procedures for field investigation and expressing my views on the use and limitations of various reconstruction techniques with an emphasis on a general methodology and the practical application of basic principles. A suggested methodology for the identification of causal factors, as well as some possible countermeasures, will be presented. I will share my opinions regarding some of the problems and deficiencies in the present practice of accident reconstruction. I recognize that some of these opinions may be unpopular, but I am convinced they need to be addressed.

Qualifications

Vehicular crashes involve dynamic interactions of complex physical systems that result in measurable damage to vehicles. In addition to damage, vehicle motion before, during and after an impact generally leaves physical evidence at the scene of the accident. The significance of this vehicular and scene evidence is not obvious to the untrained observer. Therefore, a competent analysis of the accident event is not possible without the education, training and experience necessary for the investigation, reconstruction and causal analysis process.

Accident Investigation

To be a successful accident investigator requires sufficient curiosity, motivation and discipline to learn and refine a distinct set of skills, including:

- Identification and interpretation of physical evidence.
- Understanding of the evidence and measurements needed for accident reconstruction and, in some cases, the causal analysis.
- Ability, in many cases, to qualitatively reconstruct the accident event.
- Capability to accurately identify, measure and map pertinent physical evidence, including the preparation of a clear and complete diagram.
- Photographic skills to adequately preserve the nature and location of the physical evidence for future use.
- Ability to conduct thorough and clear witness interviews.

Specific classroom education and extensive field training are required in order to achieve these skills. The normal training received in a police academy is insufficient since it is directed toward law enforcement and accident *reporting*. Simply being a licensed private investigator is also not an adequate qualification for accident investigation. There are, however, training programs and literature specifically directed to accident investigation.

Accident Reconstruction

Proper reconstruction of an accident requires the ability to collect and interpret evidence but also the ability to visualize and qualitatively recreate the accident. Also required is the ability to accurately quantify the important variables that describe the accident events. A thorough understanding of college level physics is absolutely essential since the observed and derived laws of physics explain the concepts and interrelationships of such things as distance, time, speed, acceleration, gravity, mass, weight, force, work, energy, impulse, momentum and material deformation. A brief physics review is provided in the Appendix.

An adequate background in mathematics is essential to understand and properly utilize these principles of physics. Math is the language necessary to describe these principles. This background would, at a minimum, include college level algebra, geometry, trigonometry, analytic geometry and calculus. Engineering courses such as statics, dynamics and solid mechanics that develop practical applications of physics principles are quite beneficial. Some basic knowledge of material science is also needed to understand the properties and behavior under loading of various materials such as metals, plastic, rubber, glass, concrete and asphalt. Finally, apprenticeship with a qualified and experienced professional is necessary to develop skills in field investigation procedures, the interpretation of physical evidence and the utilization of appropriate reconstruction methodology.

Many people experienced in research and consulting in this field, including myself, are aware of two categories of individuals with serious qualification deficiencies. In the first case potentially qualified persons (for example, graduate engineers) take a short course in accident reconstruction and enter the practice without any meaningful apprenticeship. They often make serious mistakes when identifying and interpreting physical evidence, leading to a general misunderstanding of the accident event. As a substitute for legitimate reconstruction methodology, they tend to misuse published formulas and computer programs. Without this relevant experience and guidance their reconstruction efforts can lead to inadequate testimony in civil and criminal litigation. These deficiencies could be addressed by serving an apprenticeship in a relevant engineering practice or research organization.

In the second more serious case, persons without the minimum educational background, such as police officers, have entered the field. They would not be considered qualified in an engineering or research setting but sometimes are allowed to testify in civil litigation. In some areas of the country this has also become a serious problem in criminal prosecution and litigation. Before I describe this problem, I should mention that my experience with police officers has, for the most part, been quite positive. Those I have had in classes were generally as enthusiastic and diligent as the engineering students I have taught. Police cooperation in our research activities was not only essential but also thorough and competent. The reports and photographs prepared by many officers are complete and very useful. The typical problem is the practice of sending police officers without the necessary educational foundation to courses for periods as short as two to six weeks. They leave with a notebook full of formulas and the belief that they are now qualified in accident reconstruction.

Digressing briefly, a good rule is to *never* use a formula that you cannot derive from basic physics principles. The derivation process, involving a specific physics concept, includes making certain assumptions that limit the applicability of the formula. If you do not understand the physics concept and the associated limitations, the chance of the proper use of the formula is quite remote. The consequences of this practice are, as expected, completely unacceptable. The use of these "reconstructions" as the basis for criminal prosecutions is in my opinion an unconscionable miscarriage of justice. It would be far more consistent with the values of our justice system to forego prosecution than to proceed with totally flawed testimony. This is not the fault of the police officer who may be unaware of his limitations and simply trying to perform his job conscientiously. The fault lies with the police agencies that follow this unacceptable practice, the prosecutors who solicit this testimony, the judges who permit it and the politicians who are unwilling to provide sufficient budgets to pay for competent expertise.

In both civil and criminal litigation the pressure and competitive drive to "win" a case can lead to the use of testimony that can range from having a subtle unintended bias to being blatantly dishonest. One possible solution to alleviate this problem, which may not be politically feasible, would be the use of court-appointed qualified experts.

Accident Causation

Determination of the cause or causes of an accident is not always obvious. It often requires special expertise from a variety of fields such as human factors, vehicle design and behavior, and highway design, maintenance and operation. Ability to properly reconstruct the sequence of accident events does not necessarily qualify one to determine all the factors that explain *why* the crash occurred. This causal analysis often requires the input from more than one analyst.

Evidence

As described in the preceding chapter, the process leading to the reconstruction and causal analysis begins with the collection and interpretation of evidence. This includes witness statements, physical evidence and available data concerning vehicle specifications, as well as roadway characteristics and geometry. We will first discuss the overrated and often misused statements of drivers and other witnesses.

Witness Statements

Driver and witness statements should not be used as the basis for any reconstruction conclusion. Evaluation of the veracity of statements is not, in my opinion, within the proper role of the accident analyst. However, most experienced individuals in this field (including myself) feel that driver statements are totally suspect due to their inherent bias.

Some common examples of these self-serving statements are:

- She hit me. I did not hit her.
- The other vehicle came out of nowhere.
- He stopped so fast I had no chance to avoid the collision.
- I was turning on red so it wasn't my fault.

- He hit me so hard, he must have been speeding.
- I only had two drinks.

Physical evidence can sometimes be used to support or totally discredit some portion of a driver's statement, but it is still the responsibility of others to judge the overall honesty of a driver's testimony.

An uninvolved witness *might* be able to provide useful information such as:

- Speed of the vehicle she was following
- Color of the traffic light he was facing
- Where the pedestrian was crossing
- Whether it was raining

Research has shown that witnesses to any event are not actually very reliable. Convictions in criminal cases based solely upon the testimony of a witness tend to give one cause for concern. People have been convicted and even executed based upon witness testimony only. Studies have shown, and most experts in related fields agree, that conclusions based upon witness testimony are a really good path to a miscarriage of justice. The reliability of the "independent witness" in an accident event has also been shown to be generally unreliable. Often a witness will observe only a small portion of the accident. From this she will, in her own mind, quickly reconstruct what she thinks must have happened and report this as an actual observation.

Someone who only saw the aftermath of a crash may give a similar statement. From this evidence he will perform his reconstruction and assume that this is what happened. Since most people are experienced drivers, they think they are experts and actually believe the assumed reconstruction to be factual.

Some common witness errors include:

- Confusing the sequence of events throughout the accident
- Grossly overestimating the number of rollovers
- Giving the wrong directions for vehicles in an intersection collision
- Estimating speed based on the noise or personal assessment of damage

Most people, including judges and jurors, substantially overestimate speed based upon vehicle damage photographs.

Another problem occurs when police, insurance adjusters and private investigators elicit answers that agree with the interviewer's preconceived opinions or desired testimony regarding the accident. The elimination of leading questions would help alleviate this problem.

The following case is an interesting example of an erroneous statement by a driver that actually proved to be useful. It was raining and a woman braked hard in an attempt to avoid a left-turning vehicle. She stated that while braking her car suddenly speeded up and collided with the other vehicle. An inspection of her car revealed no evidence of brake failure or unwanted acceleration. After some thought, this led to a careful examination of the roadway geometry. A significant depression was found in the pavement where water would have collected without draining away. This pool of water had actually caused hydroplaning. The causation of the accident was a highway defect and not a vehicle problem. In view of this finding her testimony now made sense.

While braking she was experiencing a significant deceleration that would cause her to brace herself on the steering wheel to prevent a forward movement. Upon entering the pool and hydroplaning the magnitude of the deceleration was reduced to near zero as shown in Figure 2.1(a), thus causing her body to move rearward, releasing the pressure on her arms. This was actually a positive change in acceleration or a positive jerk j where $j = da/dt$. This would create the same sensation for the driver as a sudden acceleration that is also a positive jerk as shown in Figure 2.1(b). A driver's sensation of acceleration or deceleration is a force tending to push a body backward or forward. Jerk is a change in acceleration that changes the force on the body, resulting in noticeable body movement. I learned an important lesson. Do not discount apparently ridiculous statements before giving them careful consideration.

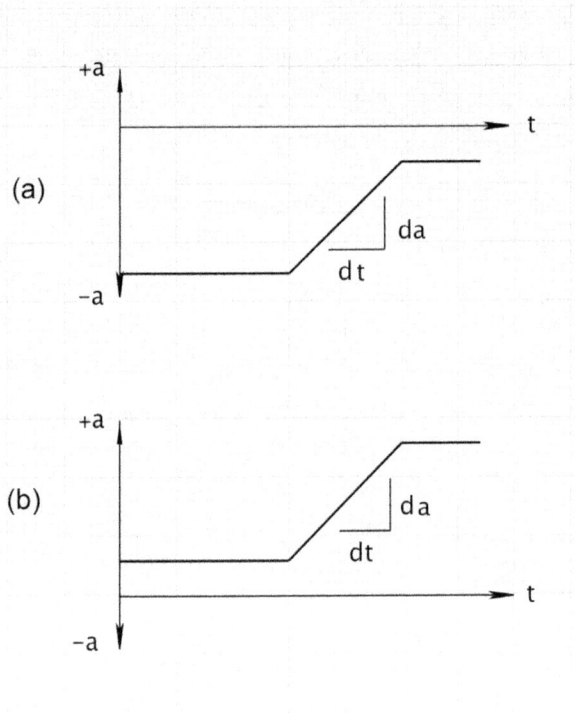

Figure 2.1. Change in Acceleration (Jerk)

Collection of Physical Evidence

Due to the proliferation of accidents and existing economic constraints, the collection of evidence is often a somewhat disjointed affair. This task generally starts with the first responders, police patrol officers, and is often the beginning and the end of any on-scene activity. The result is essentially a crash that has been reported, not investigated. Any further efforts directed toward a reconstruction are compromised by the loss of perishable evidence.

In the case of fatal or serious injuries, trained police or civilian accident investigators are often called to the scene, thus allowing for a quality field investigation. Using civilian accident investigators has often proved to be a beneficial practice since they can omit the normal law enforcement training and responsibilities and concentrate solely on accident investigation.

The following procedural steps should, in general, guide the on-scene investigation:

1. Secure the scene to prevent contamination by vehicles and bystanders.
2. Confer with first responders and, if possible, participants and witnesses.
3. Conduct an initial examination of the accident scene and the vehicle(s).
4. Form a *tentative* qualitative reconstruction of the accident events.
5. Identify all the physical evidence that *might* be important.
6. Photograph and interpret all of this evidence.
7. Locate and document the physical evidence.

The physical evidence to be identified, documented and photographed would include:

- Final rest locations of all vehicles, bodies and pertinent debris
- Pavement marks such as skid, yaw, cornering, redirection, and impact marks
- Chops, gouges, scrapes and scratches in the pavement
- Fluid trails
- Off-road tracks and furrowing
- Vehicle damage
- Highway damage and conditions that may be related to accident reconstruction and/or causation

Scene Photography

Since the quality of some scene physical evidence will quickly deteriorate, photographing this evidence should usually take priority over taking some measurements. Photographic coverage of an accident scene is not an artistic endeavor. Dramatic photographs showing bodies and rescue efforts may be desirable for the evening news but are not useful for reconstruction. This effort should be directed toward providing the analyst sufficient information to, at least, qualitatively reconstruct the accident and to provide a cross check of the field measurements and diagrams. Photographs do not, however, replace the need for accurate measurements.

A good way to start the process is to take a series of photos following the path of each vehicle from its pre-crash location to final rest. The initiation of this photo series should start well before the initial evidence to provide a broad perspective. Photos looking back at vehicle approaches are helpful in showing the pre-crash situation.

A common problem is that photos of the physical evidence consisting of close-ups only do not allow the user to locate the evidence relative to the general accident scene. Each piece of physical evidence (final rest position(s), tire marks, etc.) should appear in photos taken from more than one direction with enough permanent features or placed markers to establish its location. It is often necessary to back away to start the process.

Then each piece of physical evidence should be photographed close enough to show tread patterns, yaw mark striations, gouge directions and other features important for later analysis. This is particularly important around the area of impact since the location and orientation of each vehicle at impact as well as the approach and departure directions are absolutely essential for reconstruction.

Nighttime investigations present a photographic challenge. Flash units are generally satisfactory for close-up photos but obtaining broad views is often difficult. One technique is to use a long exposure time that will yield excellent nighttime photos but may require the use of a tripod. Finally, check the quality and coverage of each photo prior to leaving the accident scene, taking additional photos if necessary.

One relatively new development that would be ideal for on-scene photography is the camera-equipped helicopter drone. This system could take video and still photos, providing the previously described coverage that could be monitored and immediately checked by the investigator. The advantages of this method would be to:

- Provide quality and complete photo coverage
- Reduce the time required to hold the scene
- Provide a broad view aerial photograph of the entire scene
- Provide a good (even to scale) photograph of the impact area from which vehicle impact locations and orientations, as well as vehicle departure direction could be determined.
- Give a top view of each vehicle at final rest

Other potential sources are public and private security cameras and witnesses' cell phone cameras. These may have photos or videos that are useful in reconstructing the accident. Checking these sources can be time well spent.

Vehicle Photography

Photographing and inspecting the vehicle(s) can often be conducted simultaneously. In many cases it is desirable or necessary to conduct some portion of this activity after the vehicle(s) has been removed from the scene. An effort should be made, however, to check for and photograph exterior and interior evidence that will deteriorate with time or may be altered by the towing process.

A printed list of items to inspect, record and photograph, if appropriate, is a good way to avoid missing important evidence. This list would include but not be limited to:

- A complete vehicle description including the VIN
- Tires – flat, detreaded, damaged, jammed, etc.
- Lights – damage and operational status
- Windshield, backlight and side windows
- Suspension, brakes and steering
- "Frame" bending such as sidesway
- Scratches, scrapes, gouges, dents, etc.
- Paint peels and material transfers
- Damage – crush, component displacement, intrusion, etc.
- Air bag deployments
- Evidence of seat belt usage such as belt stretching and D-ring impressions
- Windshield fractures (star-shaped from interior contract)
- Side glass collapse
- Distortions of the steering wheel, gear shift or rearview mirror
- Dents and damage to the dash, doors and seat backs
- Smears, cloth impressions, hair, blood and tissue
- Evidence of ejection through windows, doors or sunroof

Vehicle photography should begin with broad views of the front, rear and both sides of the vehicle. When in doubt, back up to provide an undistorted perpendicular view. Then take closer photos parallel and perpendicular to the damaged sections of the front, rear and each side. Oblique photos are often useful in illustrating crush extent.

Close-up photos are required to show details of gouges, scrapes, dents, paint peels and material transfers. Even in daylight, flash fill can help reveal these details. Undercarriage views can show vehicle components that have struck the pavement. Top views, where possible, often provide excellent views of damage profiles.

Inspecting and photographing the interior of the vehicle is not always essential for accident reconstruction but can preserve evidence necessary for determining the occupant kinematics and the injury causation mechanisms.

Scene Measurements

Documentation of the physical evidence should start with the selection of a reference point and reference line that are, if possible, permanent in nature. The reference point can then be used as the origin for an $x - y$ coordinate system and for the location of a surveying instrument.

Virtually all scene measurements can be made using increasingly reliable and sophisticated surveying equipment. Although this method may require two people, it provides accurate measurements and limits the time needed for occupation of the accident scene. Corresponding computer programs can translate these three dimensional data into x, y and z coordinates and a useful scene diagram.

Location of scene evidence can also be obtained using something as simple as a roll-a-tape. Measurements along a reference line (for example, the x axis) and then perpendicular to the line yield the $x - y$ coordinates of the item of evidence.

An accident scene with complex roadway geometry makes it difficult to use an $x - y$ coordinate system for measurements. It may be necessary to use a triangulation system to locate vehicle final rest positions and other physical evidence. The location can be defined by any two measurements from permanent fixtures to the item in question. Preparation of a diagram will, however, be somewhat cumbersome.

Regardless of the methodology used, a list of all items of physical evidence should be completed with a brief description and their x and y coordinates. Vehicle final rests should include at least two wheel locations – vehicle rest is not a single point. The beginning and end points of tire marks, gouges, scrapes and scratches should be recorded. A field sketch should be prepared and, hopefully, later converted to a scale diagram with the pertinent evidence clearly labeled.

Some police agencies are now using a practice that is frankly astonishing. After completion of the scale diagram, the field notes and measurements are destroyed. This should disqualify a diagram as courtroom evidence since it cannot be verified. Any legitimate motivation for this practice eludes me.

The measurement of impact departure directions is essential for a meaningful reconstruction of the accident. A large protractor can sometimes be used to obtain a reasonable estimate. In post impact spinouts the only evidence may be a set of curved yaw marks that do not directly yield the direction of travel of the vehicle's center of gravity.

A simple, but somewhat cumbersome, device for tracking the path of the vehicle's center of gravity is the adjustable vehicle jig shown in Figure 2.2. This consists of a longitudinal bar (or two) with a movable center of gravity marker. The four legs can then be adjusted to conform to the *damaged* vehicle's wheelbases and track widths. Placing the wheel points on the corresponding yaw marks at a series of locations will then establish the path of the vehicle center of gravity leading from the point of impact. Moving it from point to point along the spinout trajectory assists in tracking the path of each individual tire. This is particularly useful where there are one or more yaw mark crossovers. This can be constructed from lightweight plastic piping that can be dismantled for ease of transport. I would recommend leaving the details to a competent mechanic. Departure directions may also be obtained from timely, scaled aerials using a methodology similar to that utilized for the accident scene.

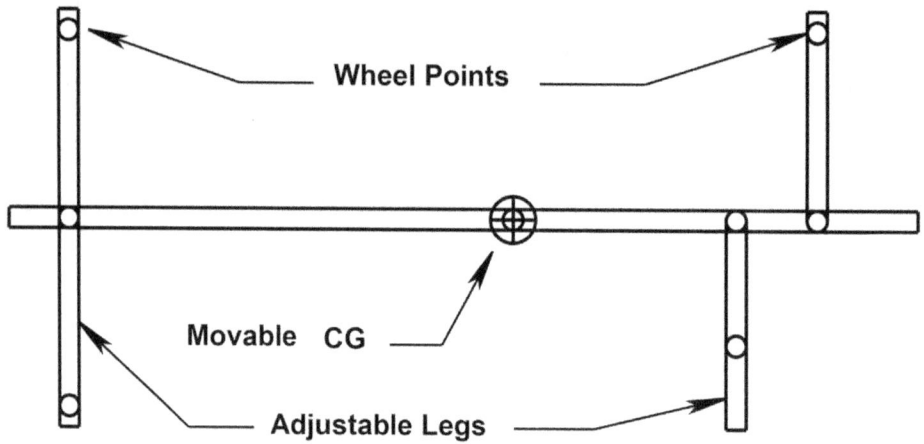

Figure 2.2. Adjustable Vehicle Jig

It is sometimes useful to determine the radius R of a highway curve. Computer programs that require three points on the curve are readily available. These points can come from the data generated by a surveying instrument or from a scaled aerial photograph. It also is fairly simple to find this radius at the accident scene by establishing and measuring a chord length .. and a middle ordinate m as shown in Figure 2.3.

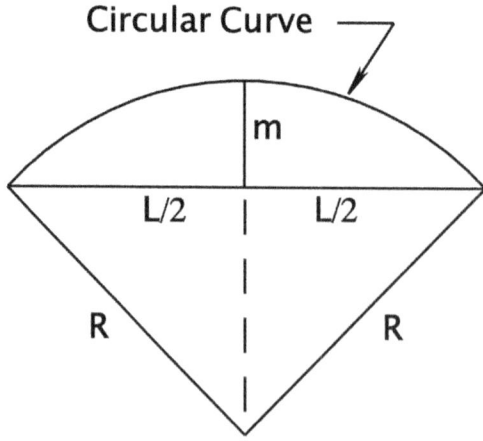

Figure 2.3. Curve Radius

From the right triangle we have the relationship

$$R^2 = (R-m)^2 + (L/2)^2.$$

Expanding yields

$$R^2 = R^2 - 2mR + m^2 + L^2/4$$

or

$$2mR = L^2/4 + m^2$$

or

$$R = \frac{L^2}{8m} + \frac{m}{2}.$$

The use of curve radius R will be discussed later.

Measurement of longitudinal slope or grade g or cross slope e or superelevation e is necessary for some reconstruction or causation analyses. These measurements may be obtained from surveying equipment or found by using a simple carpenter's level as shown in Figure 2.4. The slope or superelevation is defined as being equal to h/L or equal to $\tan\theta$.

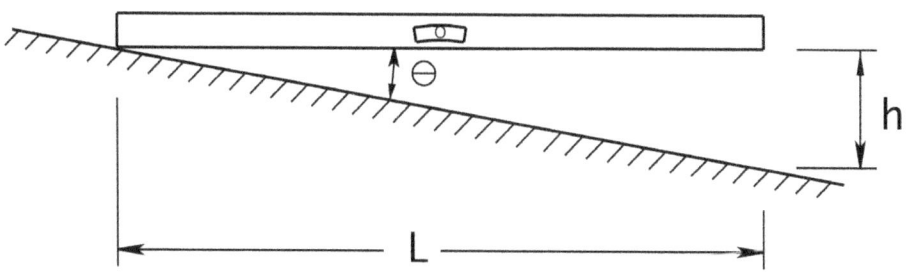

Figure 2.4. Slope

For example, a four-foot (48 inch) level with a measured height of 1 ½ inches will yield a slope g equal to

$$g = 1.5 in / 48 in = 0.031 \ in/in$$

or

$$g = 0.031 \ ft/ft$$

or expressed as a percent

$$g = 3.1\%.$$

The actual angle $\theta = arc \ \tan \ 0.031 = 1.8$ degrees.

Some accident configurations require a determination of the relative paths of different tires of the vehicle. For example, a tractor-trailer making a sharp turn might strike a pedestrian with the side of the trailer. This can result in the pedestrian being then run over by the trailer wheels since they are tracking at a smaller radius. Consider a simple four-wheel vehicle making a right turn. The front tires are following two different paths and the rear wheels are tracking inside these paths, each following a different radius. The vehicle's center of gravity will, of course, be tracking at a fifth radius.

When I first encountered this problem of determining the wheel paths, I confidently sat down and proceeded to calculate these relationships in reference to the front wheel steer angle, the track width and the wheelbase. It took awhile for me to realize that, due to the complexity of the physics involved, an empirical solution was required.

As the front tires are turned, they are subjected to lateral forces that produce side-slipping as they roll. Therefore, the actual path of these tires is outside the path indicated by the steer angle. That is, the radius of the path is greater than one might expect. There is also a problem predicting the paths of rear tires in relation to front tires. Since the rear wheels are fixed in line with the vehicle, they will be twisting and slipping as well as rolling, thus creating a complex problem. This problem becomes more important and much more complex for an 18-wheel tractor-trailer. There are five axles, four with dual wheels connected in various ways, as well as the articulation of the trailer to consider.

An experienced investigator with no math, physics or reconstruction background gave me the solution to this reconstruction problem. It requires the accident vehicle or an exemplar vehicle and some cheap paint that will quickly fade. Place a small pool of paint in front of each of the tires in question and execute the turn. The solution will be shown in living color. A good source of wheel tracking information for a variety of generic vehicles is a recent edition of the American Association of State Highway and Transportation Officials' (AASHTO) "A Guide on Geometric Design of Highways and Streets" in the section titled "Design Vehicles."

It is important to remember that, as the cornering speed increases, the rear wheels will tend to follow a path that has a radius equal to and even greater than the path of the front wheels. This off-tracking is readily apparent in the typical one-quarter mile dirt track racing events. The acceleration force from the rear wheels has a component that produces the centripetal force required to maintain the vehicles circular path. A tractor-trailer can display this off-tracking with the trailer wheels well outside the path of the tractor as shown in Figure 2.5. This is sometimes the precursor to a rollover.

Figure 2.5. Slow Yaw – Tractor Trailer

Vehicle Measurements

Forms showing a generic diagram of the type vehicle being inspected can facilitate the documentation and measurement of vehicle damage. This provides a convenient way to record damage location and measurements such as wheelbases, front and rear overhangs and damage profiles as shown in Figure 2.6. The damage beginning and end should, if possible, be referenced to some unaffected portion of the vehicle. Descriptions of gouges, scrapes, paint peels and material transfers should also include their locations.
The same techniques used for scene measurements are also applicable for obtaining a complete description of vehicle damage.

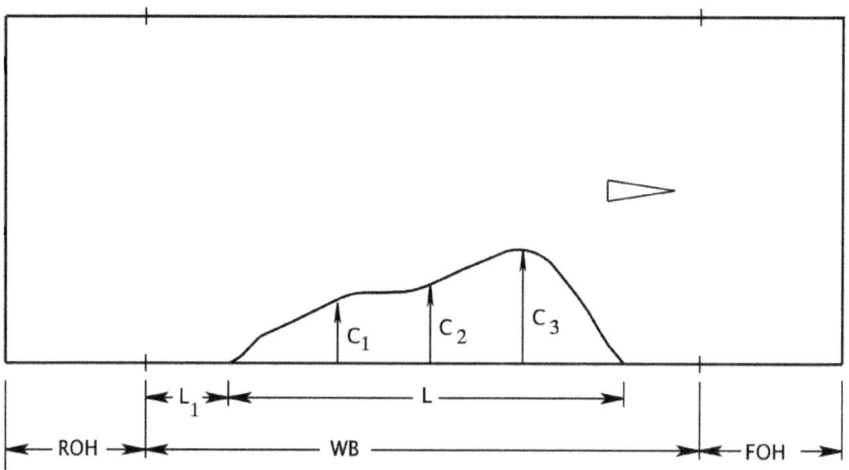

Figure 2.6. Damage Profile

Interpretation of Physical Evidence

Application of the principles of physics in the reconstruction process is totally useless without the identification and proper interpretation of the available physical evidence. An often-neglected basic piece of evidence is the physical scene. A properly scaled aerial photo or detailed scene drawing is necessary for the proper documentation of evidence and as a framework for the reconstruction.

The clues to what happened during an accident consist of evidence left on the street, the damage to and location of vehicles and bodies, and the internal evidence of occupant motions. Starting with the street evidence, we will discuss some of the important items.

Tire Marks

Tire marks show the location and motion of a vehicle. These include tire tracks, skidmarks, yaw marks, scuff marks, redirection marks, cornering marks, side scuffs, acceleration scuffs, and loading scuffs.

Tire Tracks

Tire tracks reveal the path of a vehicle at anytime during the accident sequence. They may be found in grass or dirt or moving through debris and fluids on the pavement. A heated tire may leave hot patch marks after skidding.

Skidmarks

A tire may be prevented from rolling by braking or by impact damage. Kinetic energy is dissipated in the form of heat that melts the tire surface, leaving a distinct black mark on the pavement.

The normal weight distribution and the weight transfer due to the forward pitch of the automobile during braking result in the front tire loading being considerably greater than the rear. This weight transfer is caused by the couple composed of the inertial force acting through the center of gravity and the opposing friction forces at the tire/pavement interface.

A car without ABS brakes or in the case of a malfunction will leave front tire skidmarks and often rear tire skids. Due to the heavy loading through the wheel rims the front tire marks will generally consist of two parallel dark lines. (This type of skidmark can also be the result of under-inflation.) The rear tire skidmarks will generally be lighter and will show the interior tread pattern. (This can also be the result of over-inflation.) The typical front and rear skidmarks are shown in Figure 2.7. An example of dual wheel rear skids is shown in Figure 2.8.

24 | HIGHWAY ACCIDENTS

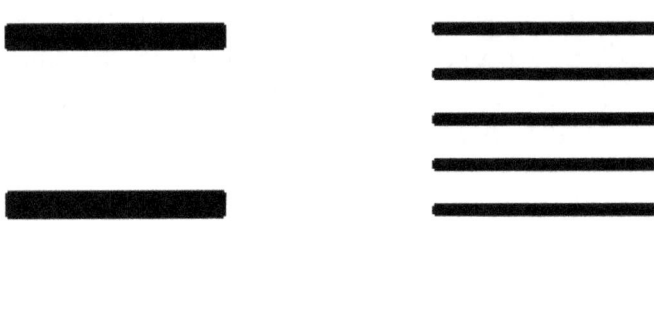

Figure 2.7. Skidmarks

Figure 2.8. Dual Wheel Rear Skids

If the vehicle is rotating during the skid, these marks will start to smear and look somewhat like yaw marks (more about this later). If one of the skidding tires is flat, it will leave dark parallel lines similar to a front tire skidmark but will generally show a distinct wobble.

In general skidmarks are not the same length. This difference can be caused by slight differences in brake adjustment or by differences in wheel loading (e.g., due to pavement cross slope). The more lightly loaded wheel will lock up first. In utilizing this evidence recognize that the braking and resulting deceleration begins slightly *prior* to the initiation of the longest skidmark. Use of the average length of skidmarks is inconsistent with the actual braking dynamics and, therefore, will yield an under-estimation of the speed loss during the skid.

Most cars are now equipped with antilock braking systems that prevent wheel lockup. There will, therefore, be little or no evidence of braking. Faint shadowing which will soon disappear may be noted in some cases.

Semis and other large trucks without functioning ABS brakes will normally leave skidmarks that are much heavier and darker. Truck tire pressures are considerably higher than automobile tires. Therefore, the friction forces per unit area, the work done and the kinetic energy dissipated will be correspondingly greater. This will cause higher tire temperatures, more molten rubber and heavier skidmarks.

There will often be several short black tire marks following the termination of the truck tire solid skidmarks. These are sometimes referred to as "skip skids." These are not skidmarks but are patches of molten rubber laid down from the hot spot on the now rotating tire. These marks can be identified by their diminishing intensity and by a spacing equal to the circumference of the tire. Actual skip skids may be deposited by the rear tires of an empty semi-trailer due to the bouncing of the under-loaded tires.

Motorcycles without ABS brakes will generally show a somewhat thinner rear skidmark only. Due to the steering required to maintain yaw stability (see Chapter 3 –Partial Braking), you often find a discernable serpentine shape to this skidmark. If both wheels are locked you will see a quick separation of the two tire marks followed by gouges and scrapes where the motorcycle has gone down (and a very unhappy rider).

Skids in dirt will generally show no tread marks but will show a definite disturbance of material. If tread marking is visible, this is most likely a rolling tire. Skids in grass will also show no tread pattern but can be identified by ripping or tearing of the grass and sod. The use of delayed infrared photography may reveal evidence of tire tracks and skids by the different temperature of dead or damaged grass.

Skidding on a wet surface results in a significantly reduced deceleration capacity. There is, therefore, less heat energy generated and much of that is dissipated in the heating of water rather than tires. Frequently there will be no discernable skidmarks. However, after drying there will often be white marks with similar tread patterns to dry skidmarks. (Think photographic negative if you are old enough to remember them.) These marks are caused by steam cleaning of the pavement surface.

It is important to remember that jammed tires will produce skidmarks regardless whether the vehicle has a functioning anti-lock braking system. In some cases a skidmark may be tied to a specific tire. Examination of the tire may show a flat spot due to rubber lost during the skid.

In cases where the direction of the skid is in question a careful (and timely) examination of the pavement may be helpful. For sharp, crisp aggregate in the pavement, rubber particles will be collected on the approach (upstream) side. For rounded aggregate this material tends to be found on the departure (downstream) side.

Yaw Marks

A rotating tire traveling off-line from the vehicle travel direction creates a yaw mark. This occurs when a vehicle suffers a loss of control or by an impact that results in a post-impact spinout. These marks are generally wider than a straight-ahead tire mark, are curved, show no tread pattern and have the characteristic striations produced by the tire rotation. (See Figure 2.9.)

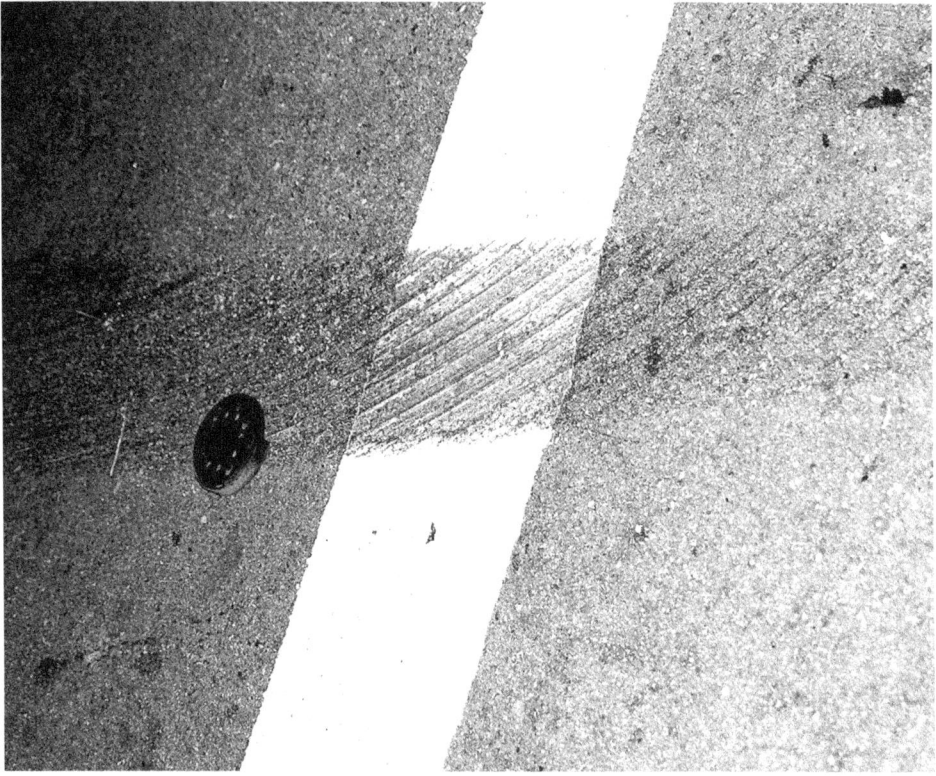

Figure 2.9. Yaw Mark

The absence of these diagonal striations may indicate that the tire is locked up and is actually a skidmark from a rotating vehicle. The radius of curvature of these yaw marks has nothing to do with "critical speed" and, therefore, *cannot* be used to estimate speed. Critical speed will be discussed later.

Cornering Marks

A vehicle traveling at a high speed or following a small radius of curvature can create a cornering mark. This tire mark, if present, is generated by the exterior edge of an outside tire. It is a thin curved black line, often quite faint. The presence or absence of a cornering mark does not indicate that the vehicle is at "critical speed" and should not be used to determine speed.

Scuff Marks

Wide, smeared marks from a tire that is being pushed sideways are generally called scuff marks. These can be helpful in locating a vehicle's position at impact.

Impact and Redirection Marks

A sudden redirection of a skid, yaw or cornering mark will place that tire's location at impact. An example is the front tire skidmark with a sudden redirection to the right as shown in Figure 2.10. A collision may also produce a short smeared tire mark due to a large downward force during impact.

Figure 2.10. Redirected Front Tire Skid

Acceleration Scuffs

A high rate of acceleration may result in tire slippage. The marks from this are generally curved and will diminish as the vehicle speed increases.

Scrapes, Chops And Gouges

Scrape marks (or scratches) in the pavement can be useful in determining departure angles and post-impact travel paths. It is important to correlate these to the vehicle parts that produced them.

Chops are rather deep marks, such as made by a garden hoe. These are produced at impact by a large downward movement of some portion of the vehicle that generally needs to be identified. A rollover will often produce a curved wheel rim mark similar but shallower than a chop mark. Gouge marks are similar to chop marks but show a definite direction. During impact a moving vehicle component is driven into the pavement. The gouge starts deep and shallows out in the direction of travel as the impact force is released.

This pavement damage can be very useful in determining the location and the movement of vehicles during and after impact.

Fluid Trails

Fluid trails between impact and final rest are also helpful in determining the vehicle's post-impact travel path. Blood smears can sometimes be used to assist in the trajectories of pedestrians and ejected vehicle occupants. Figure 2.11 shows a black impact tire mark, white gouges from the undercarriage and the beginning of a post-impact fluid trail. The termination of this fluid trail is shown in Figure 2.12.

Figure 2.11. Impact Tire Marks, Gouges and Fluid Trail

Figure 2.12. Fluid Trail

Debris

The location and pattern of debris is rarely useful in determining a good estimate of the location of the point of the collision (POC). Vehicle parts (metal, glass, plastic, etc.) drop to the street, scatter and move in the general direction of travel.

If there is no horizontal movement the pattern of scatter will be circular. When there is horizontal motion the pattern will generally be similar to that shown in Figure 2.13. If debris paths can be followed back to their point of intersection, this would indicate the point (and general direction) where the debris hit the pavement. This could be related to the POC if you knew the speed of debris post-impact and how far it dropped to the street. At best this may be used as a crude estimate only.

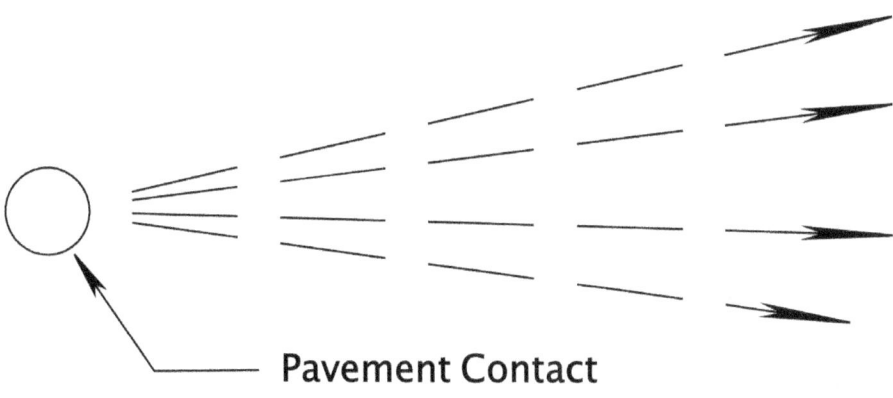

Figure 2.13. Debris Field

Vehicle Damage

The location and shape of the damage (crush profile) of each vehicle can be used to determine the relative location and orientation of the vehicles for head-on, rear-end and angle impacts. If only one vehicle's location and orientation at impact can be determined from street evidence, this damage evidence will establish the location and approximate orientation of the other vehicle.

The length and depth of the crush profile can be used to determine the loss of kinetic energy during the impact. This will be discussed later in Chapter 4 Impact Dynamics.

Gouges and scrapes are similar to those found in the pavement. They start deep and shallow out as contact is lost. They identify contact points with vehicles and other objects and indicate the relative direction of motion. Horizontal scrapes and scratches indicate slippage between a vehicle and an impacted object or between two vehicles during a collision. The frontal damage shown in Figure 2.14 is typical of a near perpendicular impact with a flat concrete barrier. Note the fine-grained horizontal scrapes indicating some small sliding at impact.

Figure 2.14. Frontal Damage

Displacement of paint (peeling), chrome, rubber or other components also determine contact points and direction of motion. The transfer of materials such as paint, concrete, dirt and vegetation provide other indications of impacts with vehicles, roadway objects and rollover contacts. Grass imbedded between the rim and tire identifies the leading tire(s) in an off-roadway spinout.

Tire air outs, blowouts and tread separations are important in assessing loss of control mechanisms. Scrapes and cuts are evidence of partial or total jamming.

Bending of the frame will frequently occur due to side impacts such as the vehicle bent around a tree as shown in Figure 2.15. Although this distortion is not included in the measurement of the side crush profile, it does contribute to the loss of kinetic energy.

Figure 2.15. Side Impact with Tree

In an intersection type of collision there will often be a lateral displacement (sidesway) of the front of the striking vehicle. (See Figure 2.16.) This bending is due to the snagging or friction between the two vehicles and is a function of the speed of the struck vehicle. Experience has shown that a significant sidesway indicates a probable speed of the struck vehicle equal to or greater than 15 mph. This is a rough guideline only and should be used with caution.

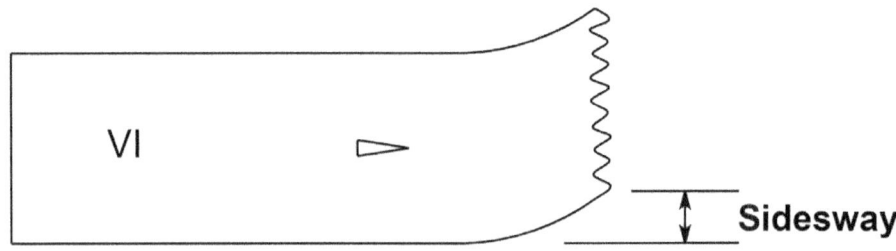

Figure 2.16. Sidesway

Interior evidence previously described may be used to indicate occupant movements since the dynamics are the same as for vehicle motion. Individuals with biomechanical or medical background should be consulted in the assessment of injury causation mechanisms and the interpretation of interior contacts.

Coefficient Of Friction

In the reconstruction of accidents it is necessary to understand the various intended and unintended movements of a vehicle. Friction forces directly affect most of these movements. The coefficient of friction f is defined as the ratio of the friction force to normal force; that is, $f = F/N$. This coefficient f is not just a function of the pavement or some other single surface. It is dependent upon the properties of both surfaces. The study of vehicle motions is primarily related to the friction forces between the vehicle tires and the pavement or roadside.

A pneumatic tire sliding across pavement dissipates kinetic energy to heat energy. This is due to two processes. The first process is due to the elastic tire surface repeatedly bending and recovering as it slides over the irregularities in the pavement surface. This visible surface irregularity or roughness is the *macrotexture* of the pavement. This flexing of the tire is also called the "hysteresis effect." The second process, due to surface friction between the tire material and the pavement surface, is primarily due to the grittiness of the pavement aggregate. This is the *microtexture* of the pavement.

The hysteresis loss in automobile tires accounts for a significant portion of the overall kinetic energy loss and is, therefore, a major contributor to the overall coefficient of friction. The much more rigid truck tire depends primarily upon surface friction and, therefore, exhibits a much lower coefficient of friction.

The coefficient of friction is also dependent upon many other factors besides the pavement macrotexture and microtexture.

- **Tire Composition:** The "rubber" compound used and the specific tread and carcass design has some effect upon the coefficient of friction.
- **Tire Pressure:** Lower tire pressure tends to increase f due to the larger tire footprint.
- **Surface Discontinuities:** Bumps, dips, manhole covers, potholes, rumble strips and other surface discontinuities tend to cause bouncing with a corresponding loss in the effective coefficient of friction.
- **Surface Contamination:** Substances such as oil (particularly when followed by the first rain), dirt, snow and ice will act as lubricants, thus reducing the friction capability.
- **Speed:** The friction factor f decreases significantly as speed increases. The tire tends to "skate" over the macro texture. Since it gets hotter during a longer skid due to a larger conversion of kinetic energy to heat energy. This then causes more rubber melting, thus lubricating the tire/pavement interface.
- **Braking Intensity:** Antilock brakes or skillful brake application with little slippage between the tire and pavement prevents tire lockup and the resulting hot spot on the tire. This, of course, increases the friction factor f.
- **Wet Pavement:** Water at the tire/pavement surface acts as a lubricant and significantly reduces the coefficient of friction. This is, of course, true for almost all surfaces.[1] This significant drop in friction factor significantly affects steering as well as braking. This is so important that highway design and maintenance criteria completely ignore the dry pavement condition.

---

[1] The coefficient of friction of silica upon silica is greatly increased by water. Interesting? Yes. Important? No.

The determination of a reasonable estimate of the coefficient of friction f is required for a quality reconstruction effort. There has been extensive testing done on a variety of vehicles, on different surfaces, at various speeds and with differing environmental conditions. There is voluminous and readily available literature describing the findings of this research. Careful judgment is, however, required in the comparison of these test results to the variables involved in the accident event under investigation.

A brief listing of coefficients of friction, often called drag factors, for various surfaces is given in Table 2.1. It is recognized that many actual surfaces will have friction factors that do not fall within the ranges given since these values are only guidelines.

Table 2.1. Coefficient of Friction for Locked Wheel Skids

	Low Speed ≤ 35 mph				High Speed ≥ 40 mph			
	Dry		Wet		Dry		Wet	
Surface	Min.	Max.	Min.	Max.	Min.	Max.	Min.	Max.
Pavement								
New	.80	1.10	.50	.80	.70	.90	.45	.75
Worn	.55	.80	.45	.65	.50	.70	.40	.60
Excess Asphalt	.50	.80	.30	.60	.40	.65	.25	.50
Unpaved	.40	.80	.40	.75	.40	.75	.40	.70
Grass	.40	.60	.30	.50	.35	.55	.25	.50
Snow	.10	.50	.25	.60	.10	.50	.25	.55
Ice	.10	.25	.05	.20	.10	.20	.05	.10

For vehicles with functioning anti-lock brakes (ABS) the friction values for most conditions will be somewhat higher. The drop-off in friction factor at higher speeds will also be less. Estimation of the coefficient of friction for ice and snow conditions is difficult to determine without careful (and hazardous) site testing. It should be noted that deep snow can actually produce a braking force. One interesting condition is a travel-polished, dusty pavement covered with a light mist. This is very similar to wet ice.

Site Testing

Another method for determining the appropriate friction coefficient is to conduct brake testing at the site. If there are locked wheel skidmarks involved, the test can involve a locked wheel skid at a known speed, or by use of an accelerometer. If your test vehicle has anti-lock brakes, then recognize that the deceleration rate shown will yield friction values somewhat higher than that experienced by the accident vehicle.

Test Vehicle: The test vehicle and tires should match the accident vehicle as closely as feasible. Using an automobile for testing related to an accident involving a truck would obviously not be appropriate.

Test Path: Tests should be done over the same pavement as the accident site.

Test Conditions: The pavement should be in the same condition as existed at the time of the accident; wet or dry with temperatures that are not vastly different.

Speed Range: The test speed should be comparable to the speed range of the vehicle involved in the accident event. If this is not possible, appropriate adjustments must be made in accordance with established speed-friction relationships.

Drag Sled: This is a surprisingly popular method to establish a value of coefficient of friction for use in a reconstruction analysis. This consists of a lightly weighted piece of tire dragged slowly across the pavement. The stiffness of the tire segment basically eliminates the hysteresis effect and therefore will yield a friction value that is much too low. The light loading of the tire diminishes the mechanical interaction between the rubber and the sharp aggregate projections (microtexture). This will then indicate a friction value that is too small.

The speed of the sled is basically a walking speed or less which would be only a fraction of the accident vehicle. This would tend to yield a friction value that is much higher than the actual values occurring at roadway speeds.

Since the conflicting deficiencies in this method do not yield remotely reliable estimates for the actual value of the coefficient of friction, they would never be used for any legitimate scientific or engineering purpose. These sled test results should never be allowed as evidence in a civil or criminal trial.

Hydroplaning

As a tire is moving through water the water tends to "pile up" in front of the tire. Drainage through the tread, out to the sides of the tire and down into the pavement can prevent or alleviate this problem. If this hydrostatic pressure builds up sufficiently it will produce an upward force in excess of the weight on the tire. This then results in a separation of the tire from the pavement with a subsequent loss of braking and steering capability. The friction factor is approximately equal to zero.[2]

Factors important to the causation of hydroplaning include the following:

- **Pavement Macrotexture**: Good macrotexture provides a path for water to escape from the tire/pavement contact.
- **Surface Porosity**: The provision of a porous surface course allows for another escape route and is an important component of pavement design and construction.
- **Tire Width**: A wide tire is more prone to hydroplaning since the water has a greater distance to travel to exit the tire/pavement contact area.
- **Tire Tread**: Lack of tread depth similarly blocks an exit path to prevent water buildup.
- **Tire Pressure**: Low tire pressure increases dry friction capabilities but increases the tendency to hydroplane since, due to the larger foot print, less hydrostatic pressure is required to produce separation.
- **Speed**: Vehicle speed is a significant factor. At higher speeds more water builds up with less time to exit the tire/pavement interface.
- **Water Depth**: As water depth increases, the propensity to hydroplane increases. Although water depth is a significant factor it should be noted that when the depth reaches a certain level it starts to produce a braking force.

Highway drainage is intended to prevent the buildup of water depth on the pavement surface. The cross slope or crown is for the purpose of draining the travel lane. Excessive cross slope has, however, a negative impact on driver control, particularly on ice or snow.

---

[2] Think of a 3000 pound water ski.

Reaction Time

An understanding of driver behavior is dependent upon many factors including how a vehicle operator responds to a potential emergency. How quickly a driver responds to a given situation is defined as reaction time and may be treated as evidence in the analysis of an accident. During this reaction interval the driver goes through a process that has often been described as PIJR – perception, intellection, judgment and reaction. However, in my opinion, an assessment of a driver's successful responses and failures is better served by a more functional description of the reaction process consisting of the following five phases:

1. Search
2. Identification
3. Evaluation
4. Decision
5. Driver action.

These will be described in some detail.

Search

This first phase involves using one's primary senses to monitor the driving task – using appropriate speed, maintaining the proper course, and being prepared to respond to any new situation. A proper search involves looking well ahead to anticipate potential decision points, looking left and right, and a routine checking of the rearview mirrors. Constant attention is required even when driving on a straight, level stretch of highway.

It should be recognized that a careful, generally attentive driver will experience momentary gaps in this search process due to the following:

- Reading road signs giving directions and information.
- Looking for street signs and addresses.
- Checking dash instruments.
- Having momentary distractions by unusual or interesting roadside events.

These unavoidable interruptions need to be considered when establishing an expected response time.

Identification

Upon seeing, hearing, or feeling "something", a driver must identify the sensory input. The time to complete this phase is often quite short especially if the nature of the input is familiar to the driver. For example, as one approaches an intersection the traffic signal turning amber is a frequent and unambiguous event. A pedestrian running out from in front of a parked truck can also be easily identified. Sometimes the identification of the sensory input is not so quickly accomplished. Some examples:

- You feel the car pulling to the right due to a steering failure. Flat tire?
- You hear a bang and feel a pull to the right.
- You hear a siren but have no idea where the sound is coming from.
- You are braking in the rain and suddenly stop decelerating. Brake failure? Hydroplaning?

Darkness, rain, fog or smoke can delay the identification of another vehicle, pedestrian or an object in the roadway. These conditions can also delay seeing and reading a stop sign, a warning sign or some other important traffic control device. Opposing traffic and background lights inhibit recognizing headlights, particularly when the approaching vehicle is a motorcycle.

Although the identification process is often quite short, certain situations can easily create delays of one-half to one second. In assessing the response of a driver this phase needs to be carefully considered.

Evaluation

Upon identifying a potential problem, the driver then has to make an evaluation of the situation to determine if it presents a threat. The time required to make this evaluation varies widely depending upon the situation and the driver's experience. Some example situations illustrate how the nature of the potential threat can alter the time required for this phase.

Let's say you come over the crest of a hill and see a tractor-trailer lying on its side perpendicular to and blocking the entire highway. It is obvious that you must react. The only decision to make is how quickly you must brake.

A vehicle making a left turn in front of you raises the question – are we on a collision course? A mathematical analysis would require a good estimate of the other vehicle's speed and path, your speed, and the distance you are from the potential collision point. With this knowledge you can then compute the time for the turning vehicle to clear your path and the time for you to arrive at the point of conflict. Comparison of these two times then yields the solution. What you actually do is use the computer between your ears that has been programmed by all of your prior experience. This analog computer does not use any numerical values for speed, time or distance, but does arrive at a quite reliable solution in a fraction of the time we would need to calculate a similar solution.

A rather common situation is created by the driver who properly stops prior to entering the roadway, but then moves forward and stops again one or more times. We are familiar with these false starts and they condition us to not immediately react to the first movement. It is clear that this situation indicates a reasonable expectation for reaction time that should be increased by at least one-half second and possibly more because of an extended evaluation phase.

Decision

Having determined that the situation does indeed present a threat, the driver must then make a decision regarding what action to take. Possible responses include the following:

- Do nothing and hope for the best
- Sound your horn
- Flash your lights
- Steer
- Accelerate
- Brake
- Some combination of the above

The time required to arrive at a decision depends upon the complexity of the situation, the number of potential responses and the driver's personal experience. Here are some time-consuming considerations that might arise:

- If I accelerate, can I clear the problem?
- Can I stop prior to the potential POC?

- Can I avoid this object by steering?
- Is a crash inevitable?
- Will braking reduce the crash severity?

Assessing these various options will, of course, extend the time involved in the decision phase. Indecision and hesitation are a frequent problem with young or inexperienced drivers.

Driver Action

Once a decision has been made, executing the maneuver(s) requires some elementary driving skills. Experience is a valuable asset, particularly if it includes prior emergencies of a similar nature. Steering or accelerating can be initiated quickly since little physical movement is required, but braking requires a movement of the foot from the accelerator to the brake pedal. This delay has been demonstrated in pre-ABS vehicles by locked wheel skidmarks at an angle due to the steering input *preceding* the braking.

Total Reaction Time

Many laboratory experiments and field tests have been conducted to assess driver reaction time. There is, however, a continuing problem in extrapolating these tests to the non-alert and/or imperfect driver. The general consensus is that the reaction time for most drivers in a simple situation will fall within the range of 0.5 to 1.5 seconds with an average value of 0.9 or approximately one second. This range for a simple reaction time roughly represents the 10th to 90th percentile. It should be emphasized that this is not just a distinction between quick and slow reactors, but also indicates that a given driver will have a significant variation in reaction time. Some drivers routinely, and most drivers occasionally, will exceed this upper value. We must remember that legally licensed drivers include the young, the inexperienced, the elderly, and those who may have physical or mental impairments.

Given the empirical data it appears that a reasonable expectation for most drivers is approximately 1.5 seconds in a simple situation. Any complexity in the evaluation phase or if the decision phase has multiple options, should raise this expectation to at least 2.0 to 2.5 seconds. Highway design standards are based upon a reaction time of 2.5 seconds with the recognition that this does not include all drivers, at all times, and in all situations.

Vehicle Dynamics 3

Upon completion of the collection and interpretation of physical evidence, a reconstruction of the accident events can proceed. These events generally consist of the pre-crash, the impact and the post-impact phases, each of which must be analyzed separately. The analysis of each phase requires the application of physics principles to describe the various vehicle motions.

This chapter is directed toward understanding the vehicle motions during the pre-crash and post-impact phases of the accident. These motions include steering, acceleration and deceleration, all of which are a function of the vehicle's interaction with the ground.

Prior to starting our discussion of vehicle dynamics, it is necessary to explain some conventions regarding vector and scalar quantities. A vector has direction, as well as magnitude, whereas a scalar quantity has magnitude only. Vector addition and scalar addition are, therefore, different mathematical computations. This difference will be noted as we proceed.

In general velocity V is a vector and its magnitude is speed S. Due to our strange system of measurements, a convention also commonly used is to express speed as V in feet/second (ft/sec) and as S in miles per hour (mph). These conflicting conventions are used throughout this book, but hopefully will not create any substantial confusion.

Steering

A steering input will place a vehicle on a curved path with a particular radius of curvature at any point in time. The centripetal acceleration required to maintain this path is given by

$$a = V^2/R \qquad (3.1)$$

where

a = centripetal acceleration in ft/sec²
V = tangential velocity in ft/sec
R = instantaneous path radius in ft.

The lateral force F_L required to produce this lateral acceleration is equal to

$$F_L = ma = \frac{W}{g}a$$

where

F_L = lateral force in pounds (lb)
m = vehicle mass in lbsec²/ft
W = vehicle weight in lbs
g = acceleration of gravity (32.2 ft/sec²).

or

$$a = F_L g / w. \qquad (3.2)$$

This friction force F_L is also given by

$$F_L = f_L W \qquad (3.3)$$

where

f_L = lateral coefficient of friction.

Substituting (3.3) into (3.2) yields

$$a = f_L W g / W$$

or

$$a = f_L g. \qquad (3.4)$$

Equating (3.1) and (3.4) yields

$$\frac{V^2}{R} = f_L g$$

or

$$V^2 = f_L g R.$$

Converting the magnitude of the velocity V in ft/sec to the speed in mph yields

$$(1.47 S)^2 = 32.2 f_L R$$

or

$$S^2 = 15 f_L R$$

or

$$S = 3.87 \sqrt{f_L R}$$

where S is the maximum speed that can be maintained upon this curve.

Roadway geometry also affects this cornering movement. The banking of the curve is called superelevation where e is the cross slope of the pavement in percent. This slope e is equal to

$$e = \tan \theta$$

where θ is the angle of the cross slope.

Therefore, the effective friction factor is $f_e = f_L + e$. If the normal crown of the roadway is maintained throughout the curve, e will be negative and $f_e = f_L - e$.

The general relationships are then given by

$$S^2 = 15(f_L \pm e)R$$

or

$$S = 3.87\sqrt{(f_L \pm e)R}.$$

A lateral acceleration a that exceeds $0.2\,g$'s or, since $a = f_L g$, exceeds ($f_L = 0.20$) becomes uncomfortable for most people. Even on dry roads it is difficult for most drivers to maintain a lateral acceleration above $0.5g$'s.

Acceleration

The maximum longitudinal acceleration rate for an automobile is dependent upon the available coefficient of friction, the engine power and the number of drive wheels. For an all wheel drive (AWD) vehicle with unlimited power the maximum acceleration is

$$a_{max} = fg = 32.2f.$$

A front or rear wheel drive only car with unlimited power would have a maximum rate dependent upon the normal front to back weight distribution and the height of the center of gravity. For most automobiles this would be approximate one-half of that for an AWD vehicle.

Most cars have insufficient power to reach this potential, even at low speeds. For most drivers a "normal" acceleration rate from a stop is approximately 4.5 to 5.0 ft/sec². A "hard" rate would be approximately twice that value.

Acceleration rates for motorcycles are generally higher due to the easy weight shift to the rear drive wheel. The acceleration rates for trucks are substantially lower due to the higher weight to power ratio. For example, a fully loaded tractor trailer combination (semi) has a value of $a_{max} \approx 1.5$ to 2.0 ft/sec² at low speeds. Since this is such a low value it is also the "normal" rate.

Acceleration capability for nearly all types of vehicles drops off dramatically, asymptotically approaching zero, as speed increases. There is extensive and readily available test data for a variety of vehicles at different speeds. Typical values of vehicle accelerations, given in Table 3.1, are approximate values only that can vary depending upon the specific vehicle and driver. The effect of a longitudinal slope can be included by adding or subtracting the acceleration due to the grade g. That is, a (corrected) = a (given) $\pm$ $32.2g$. Note that is this case g is grade, not the acceleration of gravity.

Table 3.1. Vehicle Acceleration Values in Ft/Sec²

Vehicle	Speed Range (mph)					
	≤ 20 mph		± 30 mph		≥ 45*	
	Min	Max	Min	Max	Min	Max
Passenger Cars						
Normal Acceleration	4.5	5.0	3.0	3.5	1.5	2.0
Hard Acceleration	9.0	10.0	4.5	5.0	3.0	3.5
Medium Trucks	3.0	3.5	1.5	2.0	0.8	1.2
Large Trucks	1.5	2.0	0.8	1.2	0.2	0.4

* All acceleration rates rapidly going to zero as vehicle approaches maximum Speed.

Deceleration

Vehicle deceleration occurs due to unintended events as well as conscious driver actions. These deceleration mechanisms include: air resistance, tire rolling resistance, engine drag and braking, as well as rollovers and spinouts due to collisions or driver loss of control.

Air Resistance

In general air resistance can be ignored except for very high speeds or when traveling into a strong head wind. High cross winds can, however, create serious problems for high speed vehicles, particularly vans, trucks and motorcycles.

Rolling Resistance

The tread of a rotating tire travels from a circular path to a horizontal path as it passes through the footprint at the tire/pavement interface. The rolling resistance or drag depends upon the material and design of the tire, the travel speed and the inflation pressure. A normally inflated tire at slow speed will have an effective drag factor of approximately $f_e \approx 0.01$. As the inflation pressure decreases this value will increase to a drag factor similar to light braking.

Engine Drag

Engine resistance and friction throughout the entire power delivery system will vary substantially from vehicle to vehicle. Manual transmissions generally offer more resistance than automatics. A stalled engine creates more drag than an engine still running. In fact an engine that is still running at slow speed can apply power rather than drag.

In estimating the deceleration effect of engine drag one must remember that this drag is transmitted through the drive wheels only. This must be correlated with any wheels that are jammed or free rolling. Motorcycles have a much higher engine drag that allows for moderate decelerations without brake application. Some typical deceleration drag factors for passenger car tires are shown in Table 3.2. The estimated values given for engine drag on a tire may be different for all-wheel drive (AWD), hybrid and electric passenger cars.

Table 3.2. Drag Factors for Passenger Cars

Condition	Speed in MPH	Drag Factor $f=a/g$
Free Rolling Tire or Car Out of Gear	≤ 20	.01
	$30 \pm$	.02
	≤ 45	.04
Tire Subject to Engine Braking Only	> 20	.08
	$30 \pm$	.10
	± 45	.16
Tire with Light Braking		.10

Condition	Speed in MPH	Drag Factor $f=a/g$
Tire with Normal Braking		.2 to .4
Tire Jammed or Locked		f_{max}

Braking

Brake application slows or stops the rotation of the tires, thus producing the desired deceleration. Normal braking maneuvers will naturally vary from driver to driver but generally yield a deceleration rate of 0.2 to 0.4 g's or an effective drag coefficient of f_e = 0.20 to 0.40.

As the braking force is increased there will be some tire spin down or slippage until a maximum effective coefficient is reached. Skillful braking is required to maintain this maximum braking efficiency since additional braking effort will result in tire lockup.

In the simple case of a locked wheel skid to a stop, the kinetic energy will be dissipated through heat at the tire/pavement interface. This kinetic energy loss is equal to the work done by the friction force F acting through the skid distance d.

This friction force is $F = fW$ where f is the coefficient of friction and W is the weight of the vehicle. The work done is therefore

$$\text{Work} = Fd = fWd. \tag{3.5}$$

The change in kinetic energy during the skid to a stop is

$$KE = \frac{1}{2}mV^2 = \frac{W}{2g}V^2 \tag{3.6}$$

where

W = vehicle weight
V = velocity at the beginning of the skid.

Equating (3.5) and (3.6) yields

$$\frac{WV^2}{2g} = fWd$$

or

$$V^2 = 2gfd$$

or

$$V = \sqrt{2gfd}$$

which is the simple skid formula with

$V =$ velocity in ft/sec
$d =$ skid distance in ft.

Converting the magnitude of the velocity V to speed S in mph yields

$$(1.47S)^2 = 2(32.2)fd$$

$$S^2 = 29.9\,fd\,.$$

This can be rounded off without any significant error to

$$S^2 = 30\,fd$$

or

$$f = S^2/30d$$

or

$$d = S^2/30f$$

or

$$S = 5.47\sqrt{fd}\,.$$

In general it is more useful to consider losses in kinetic energy rather than losses in speed. For our purposes S^2 may be considered to represent the energy lost during a skid.[3] For example if the skid occurs over more than one surface, the total energy loss S_T^2 is the sum of the individual losses over each surface.

That is

$$S_T^2 = S_1^2 + S_2^2 + ...$$

or

$$S_T^2 = 30(f_1 d_1 + f_2 d_2 + ...)$$

or

$$S = 5.47\sqrt{f_1 d_1 + f_2 d_2 + ...}\ .$$

This is often referred to as the combined speed skid formula. Note that the following expression is *not* true

$$S_T \ne S_1 + S_2 ....\ .$$

If braking occurs on a longitudinal slope, the effective coefficient of friction is equal to

$$f_e = f \pm g$$

where g is the grade of the slope in percent. That is,

$$g = \tan\theta$$

where θ is the angle of the longitudinal slope.

---

[3] I recognize that this is not actually correct but no error will be introduced if you have a constant weight throughout the motion of the vehicle.

Partial Braking

A vehicle that has braking applied to all wheels will decelerate at a rate dependent upon the coefficient of friction only. If all wheels are not involved, the braking efficiency will also be a function of the weight distribution during the braking maneuver. This weight distribution is dependent upon the horizontal and vertical location of the vehicle's center of gravity. This information is available for most passenger cars.

If the location of the center of gravity (CG) of a two-axle vehicle is not known, it can be determined using the test shown in Figure 3.1. With the vehicle level the horizontal position of the CG can be found by independently weighing the front and rear axles.

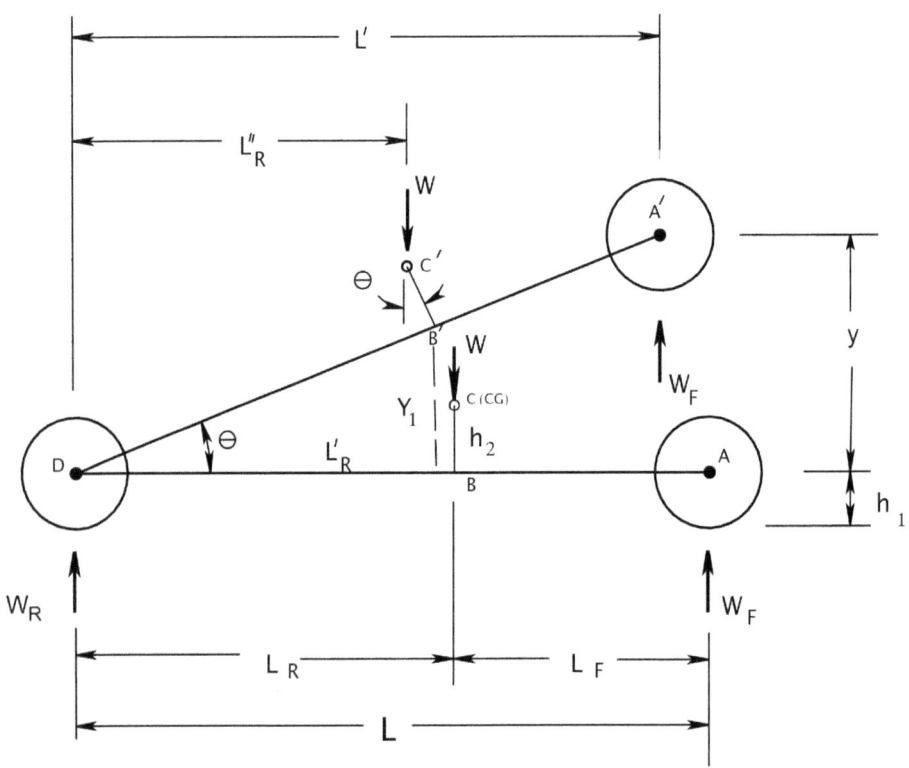

Figure 3.1. Locating Center of Gravity

From static equilibrium, summing moments about point D yields

$$\sum M_D = W_F[L] - W[L_R] = 0$$

or

$$L_R = \frac{W_F}{W} L \quad (3.7)$$

and

$$L_F = L - L_R$$

and

$$W = W_F + W_R$$

where

L = wheelbase
L_F = distance to the CG from the front axle
L_R = distance to the CG from the rear axle
W_F = load on the front axle
W_R = load on the rear axle
W = vehicle weight.

The vertical position of the CG can then be found by raising the front axle a known distance y and weighing the front axle again to obtain the new value of the load on the front axle W'_F.

The vertical dimensions of interest are

y = height that the front axle was raised
h_1 = original height of the axle above ground
h_2 = height of the CG above the axle
h = height of the CG above the ground
$h = h_1 + h_2$.

The angle through which the wheelbase has been rotated is equal to θ. Other horizontal measurements involved are

L' = distance from point D to A'
L'_R = distance from point D to B'
L''_R = distance from point D to C'
x_2 = lateral displacement of point C' relative to B'
$x_2 = L'_R - L''_R$ (not shown on Figure 3.1).

The angle of rotation θ and $\tan\theta$ are found from the expression

$$\sin\theta = y/L.$$

The pertinent relationships are then given by

$x_2 = h_2 \sin\theta$
$L' = L\cos\theta$
$L'_R = L_R \cos\theta$
$L''_R = L'_R - x_2$
$L''_R = L_R \cos\theta - h_2 \sin\theta.$

Determination of the relationship between W'_F and h_2 can be found by considering the new equilibrium condition. Summing moments about point D yields

$$\sum M_D = W'_F[L'] - W[L''_R] = 0$$

or

$$\frac{W'_F}{W} = \frac{L''_R}{L'}$$

or

$$\frac{W'_F}{W} = \frac{L_R \cos\theta - h_2 \sin\theta}{L\cos\theta}.$$

Since

$$\sin\theta / \cos\theta = \tan\theta$$

$$\frac{W'_F}{W} = \frac{L_R}{L} - \frac{h_2}{L}\tan\theta. \tag{3.8}$$

Rearranging to find h_2 yields

$$\frac{h_2}{L}\tan\theta = \frac{L_R}{L} - \frac{W'_F}{W}$$

$$h_2 = \frac{L}{\tan\theta}\left(\frac{L_R}{L} - \frac{W'_F}{W}\right). \tag{3.9}$$

Using an example vehicle with

$L = 10$ ft
$h_1 = 1.5$ ft
$W_F = 3000$ lb
$W_R = 2000$ lb
$W = 5000$ lb.

From Equation (3.7) we have

$$L_R = \frac{W_F}{W}L$$

$$L_R = \frac{3000}{5000}(10) = 6.0 \text{ ft}$$

$$L_F = 10.0 - 6.0 = 4.0 \text{ ft}.$$

As an example, let us assume that the front axle has been raised to a vertical distance $y = 3.0$ ft and the new weight on the front axle is $W'_F = 2686$ lb. From geometry

$$\sin\theta = \frac{y}{L} = \frac{3}{10} = 0.30$$

and

$$\theta = 17.46°$$

and

$$\tan\theta = 0.3145.$$

Substituting into Equation (3.9) yields

$$h_2 = \frac{10.0}{.3145}\left(\frac{6}{10} - \frac{2686}{5000}\right) = 2.00 \text{ ft}$$

and

$$h = 2.00 + 1.50 = 3.50 \text{ ft}.$$

A reverse example is to assume that $h_2 = 2.00$ ft and we solve for W'_F. Substituting in Equation (3.8) yields

$$\frac{W'_F}{W} = \frac{6.0}{10.0}(.3145) = 0.5371$$

or

$$W'_F = 0.5371(5000) = 2686 \text{ lb}.$$

This provides a cross check on Equations (3.8) and (3.9).

Some caution should be exercised in the use of Equation (3.9) for the determination of the height of the CG. The accuracy of the scale used is important. In our previous example an error of only 1% will affect the computation as follows

$$W'_F = 2686(0.99) = 2659 \text{ lb}$$

$$W'_F / W = 2659 / 5000 = 0.5318.$$

From Equation (3.9)

$$h_2 = \frac{10}{.3145}\left(\frac{6}{10} - .5318\right) = 2.17 \text{ ft}$$

$$h = 2.17 + 1.50 = 3.67 \text{ vs } 3.50 \text{ ft}.$$

The error in the value of h is $0.17/3.50 = 0.049 \approx 5\%$.

It is, therefore, obvious that the sensitivity to inaccuracies in the weighing process is significant.

If we have the data for the location of the center of gravity, then various braking situations can be addressed. For this analysis we will use a two-axle van as shown in Figure 3.2, with the following dimensions: $L = 10.0$ ft, $L_F = 4.0$ ft, $L_R = 6.0$ and $h = 3.5$ ft.

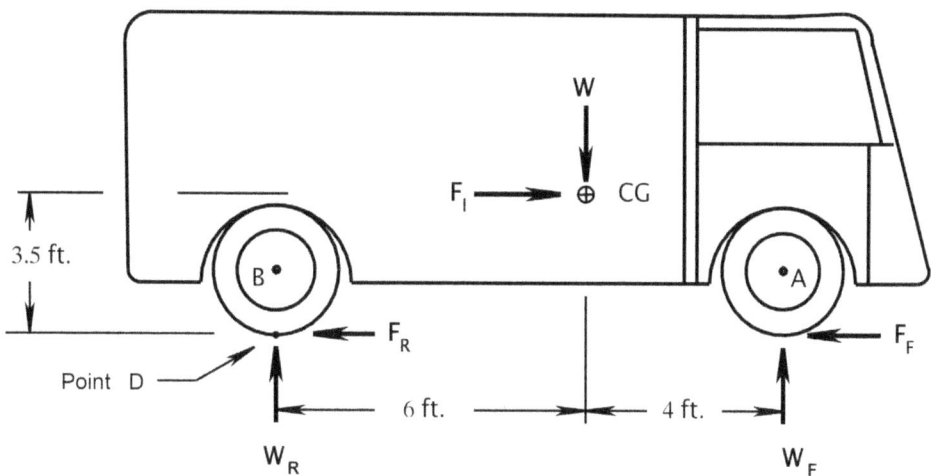

Figure 3.2. Partial Braking

When no braking is taking place, the load will be distributed in accordance with Equation (3.7), that is, $W_F = 0.6W$ and $W_R = 0.4W$. With full braking there will be a load shift to the front axle dependent upon the actual maximum coefficient of friction. For equilibrium, summing moments about point D

$$\sum M_D = W_F[10] - W[6] - F_I[3.5] = 0$$

or

$$W_F(10) = W(6) + F_I(3.5) \qquad (3.10)$$

where F_I is the inertial force

$$F_I = \frac{W}{g}a$$

and a is the deceleration rate $a = fg$. Summing force in the x direction

$$\sum F_X = F_I - (F_F + F_R) = 0$$

or

$$F_I = F_F + F_R.$$

Since $(F_F + F_R)$ is equal to fW, Equation (3.10) is reduced to

$$W_F(10) = W(6) + fW(3.5).$$

For a coefficient of friction $f = 0.5$, W_F is given by

$$W_F = (6W + 1.75W)10$$

or

$$W_F = 0.775W$$

and

$$W_R = 0/225W.$$

This explains the difference in the normal appearance between front and rear tire skidmarks. The braking efficiency is, however, not affected by this load shift since the effective coefficient of friction $f_e = f \approx 0.5$.

If only the front brakes are working (or the front tires are jammed), $F_R = 0$ and Equation (3.10) reduces to

$$W_F(10) = W(6) + fW_F(3.5).$$

For a coefficient of friction $f = 0.5$

$$W_F(10) = W(6) + 0.5W_F(3.5)$$

or

$$W_F(10 - 1.75) = W6$$

or

$$W_F = 6/8.25 = 0.73W.$$

Therefore, the effective coefficient of friction f_e would be equal to

$$f_e = 0.73f = .73(.50) = 0.36.$$

This would be a braking efficiency of 73%.

For a coefficient of friction $f = 0.20$ (light braking), without showing the calculations, the effective coefficient of friction is equal to

$$f_e = 0.65W \quad (65\% \text{ efficiency}).$$

It can be shown that for lower friction values, there will be less load transfer and the braking efficiency will approach 60%. This is, of course, the loading percentage on the front wheels with no braking.

For rear wheel braking only it can be shown in the same manner that

If $f = 1.0$, $f_e = 0.40f$.

If $f = 0.5$, $f_e = 0.44f$.

If $f \approx 0$, $f_e \approx 0.50f$.

Note that these calculated values are for the specific vehicle shown in Figure 3.2.

In the preceding discussion it has been tacitly assumed that the vehicle was a rigid body. During braking the deformation of the front tires and the existence of a non-rigid suspension system will allow a forward pitching motion that will cause the center of gravity to move forward. This will transfer additional loading from the rear wheels to the front wheels.

Therefore, the actual effective coefficients of friction are greater for front wheel braking only and less for rear wheel braking only.

A useful expression for the effective coefficient of friction is given by $f_e = pf$ where p is the percent efficiency. For most passenger cars under normal roadway conditions values for p are

$p \simeq$ 55-75% for front wheel braking only

$p \simeq$ 30-45% for rear wheel braking only.

For motorcycles the determination of the location of the CG is much more complicated. There is a large variation in the engine location relative to the front and rear wheels, as well as the horizontal and vertical location of the driver's seat. Unlike automobiles the weight and posture of the rider greatly affects both the vertical and horizontal location of the composite rider/motorcycle CG.

In general the braking efficiency for motorcycles without ABS brakes is equal to:

$p =$ 30-45% for rear wheel lock up only

$p =$ 40-85% for rear wheel lock up plus front braking

$p =$ 100% for both wheels locked. This, however, will last only a short time since stability will be lost and the motorcycle will not remain upright. It is obvious that, contrary to popular belief, motorcycles without ABS brakes cannot skid to a stop as efficiently as a passenger car.

For trucks, including tractor-trailers, the approach to the analysis of partial braking is the same except more cumbersome. It is generally further complicated by inclusion of the cargo in the determination of the CG location.

Another example of partial braking is the case of a vehicle towing a trailer without brakes. The solution consists of two steps. First, it is necessary to determine the portion of the trailer's weight that is being carried by the towing vehicle's trailer hitch as shown in Figure 3.3.

VEHICLE DYNAMICS | 61

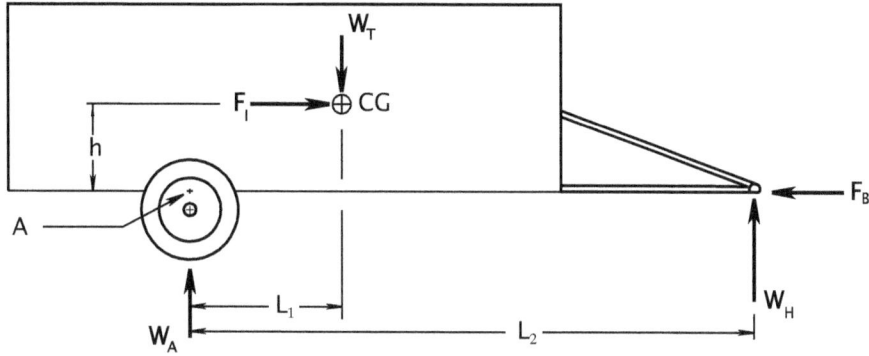

Figure 3.3. Trailer Without Brakes

For equilibrium, summing moments about point A yields

$$\sum M_A = W_H[L_2] - F_I[h] - W_T[L_1] = 0$$

or

$$W_H L_2 = F_I h + W_T L_1. \tag{3.11}$$

Summing forces in the x direction yields

$$\sum F_X = F_I - F_B = 0$$

or

$$F_I = F_B. \tag{3.12}$$

Substituting (3.12) into (3.11) yields

$$W_H = \frac{F_B h + W_T L_1}{L_2} \tag{3.13}$$

where

W_A = weight on trailer axle
W_H = weight on trailer hitch
W_T = total weight of trailer

F_B = braking force from towing vehicle
F_I = inertial force on trailer due to braking
L_1, L_2 and h are as shown in Figure 3.3.

The braking force exerted by the towing vehicle is equal to

$$F_B = (W_H + W_V)f \qquad (3.14)$$

where

W_V = weight of towing vehicle
f = coefficient of friction.

Substituting (3.14) into (3.13) yields

$$W_H = \frac{(W_H + W_V)fh + W_T L_1}{L_2}$$

or

$$W_H - \frac{W_H fh}{L_2} = \frac{W_V fh + W_T L_1}{L_2}$$

or

$$W_H \left(1 - \frac{fh}{L_2}\right) = \frac{W_V fh + W_T L_1}{L_2}$$

or

$$W_H = \left(\frac{W_V fh + W_T L_1}{L_2}\right) \div \left(1 - \frac{fh}{L_2}\right). \qquad (3.15)$$

Although this is somewhat cumbersome, if the values for W_T, W_V, f, L_1 and L_2 are known and h, the height of the trailer center of gravity, can be determined as previously described. This permits a direct solution of the trailer hitch weight W_H.

The total weight of the towing vehicle is, therefore,, equal to $W_V + W_H$ and the effective coefficient of friction is given by

$$f_e = \left(\frac{W_V + W_H}{W_V + W_T}\right) f.$$

If h and L_1 are small and W_T / W_V is small, the solution may be simplified to

$$f_e = \left(\frac{W_V}{W_V + W_T}\right) f.$$

If the trailer is heavy, this would be the case of the tail wagging the dog, with a strong possibility of loss of control.

Anti-Lock Brakes

As brake pedal pressure increases, the deceleration rate increases up to a maximum value. At this point there is some spin down of the tire rotational speed. Therefore, there is now some slippage between the tire and the pavement. It is possible, but extremely difficult, to achieve this optimum braking condition, particularly in a panic avoidance situation.

An extremely important innovation has been the anti-lock braking system (ABS). As in optimum braking this diverts much of the heat energy to the brakes, thus limiting tire heat-up and the corresponding reduction in the friction factor.

This ABS system not only provides some improvement in actual braking efficiency (increased f) in emergencies, but also prevents the loss of steering capability. Most importantly, this eliminates the need for any special effort or skill on the part of the driver. There is one drawback for the accident analyst. The absence of pre-crash panic skid marks makes it difficult to verify the approach direction and to compute the pre-braking speed. Our loss, however, is a great gain for the safety of the driving public. For motorcycles ABS braking provides a more significant increase in braking efficiency, eliminates the loss of steering and, most importantly, prevents the loss of stability.

Braking and Steering

Braking and steering can occur simultaneously if the total friction demand does not exceed the capability of the tire/pavement contact surface. As shown in Figure 3.4, the force F_R is the vector sum of the longitudinal braking force F_B and the lateral steering force F_L.

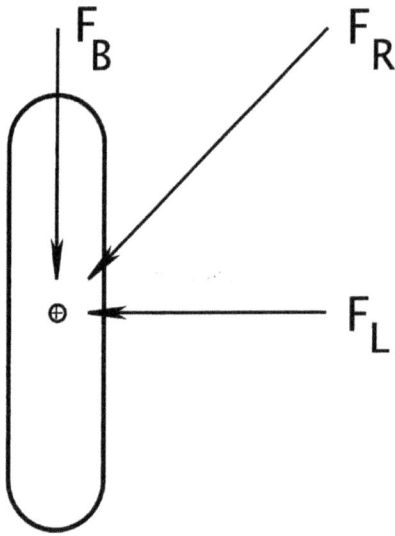

Figure 3.4. Tire Forces – Braking and Steering

That is

$$\vec{F}_R = \vec{F}_B + \vec{F}_L \text{ (vector sum)}.$$

Since all of these friction forces are a function of the same tire loading, the total friction coefficient demand is given by:

$$\vec{f}_T = \vec{f}_B + \vec{f}_L \text{ (vector sum)}.$$

If this total friction demand f_T exceeds the available friction, then steering or braking must be reduced to maintain control. Since steering can only occur if the tire is rolling, front wheel lockup will cause a total loss of steering.

Partial braking, not only reduces braking efficiency, but also can affect the vehicle stability, as shown in Figure 3.5. For the sake of simplicity, a two-wheel vehicle is used to show the stable and unstable conditions.

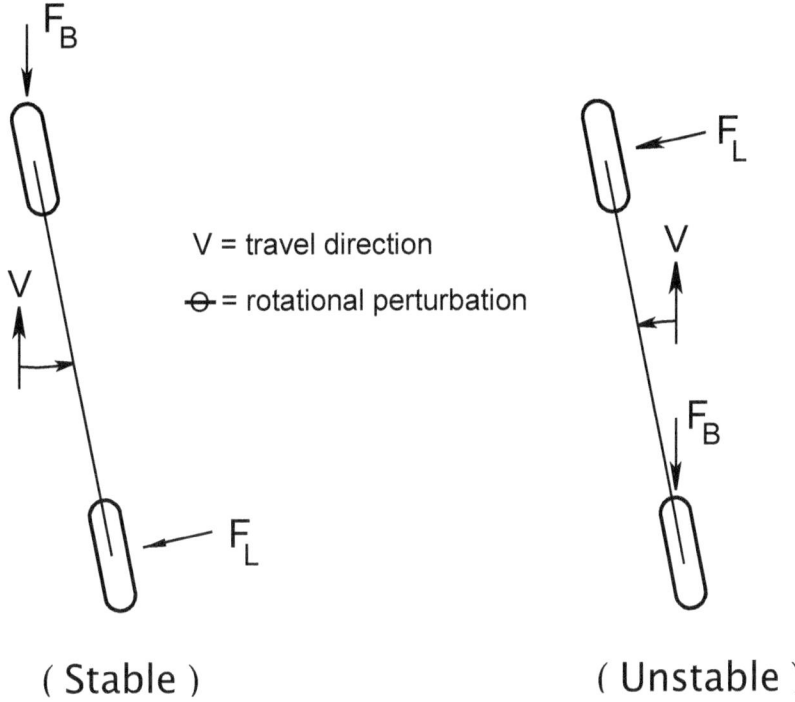

Figure 3.5. Partial Braking – Stability

A stable condition occurs when only the front wheels are locked. The front tire is skidding. Therefore, the braking force is directly opposite to the travel direction. Since the rear wheel is rolling freely, any small rotational perturbation will produce a lateral force F_L that will restore the vehicle to its proper alignment. A small perturbation can be caused by a variety of things, including a steering input or a pavement cross slope.

With a rear wheel lockup only, the braking force on the non-rotating rear tire will be in line and opposite to the direction of travel. Any small perturbation will, however, create a lateral force F_L on the free-rolling front tire that will increase the rotation, thus leading to instability. (See Figure 3.5.) Without a precise steering correction the vehicle will rotate 180 degrees into the previously described stable figuration.

This inherent directional instability that occurs with rear wheel lockup not only causes steering problems for all vehicles, but also creates vertical stability problems for motorcycles. In car chase movie scenes this is the simple way to create a reversal of direction. Lockup of the rear wheels with a quick steering input and you have a quick U-turn. But, as the old saying advises, "Don't try this at home." In an all-wheel skid rotation of the vehicle may still take place due to some pre-skid steering maneuver. The deceleration rate will, however, stay essentially the same as a straight-ahead skid.

Slow Yaw

A loss of control caused, for example, by hitting a patch of ice often results in a slow yaw. This rotation of the vehicle about the vertical Z axis will produce a deceleration due to the changing angle of the tires relative to the vehicle velocity vector. As shown in Figure 3.6, the effective drag factor for a free-rolling tire will vary from zero at $\theta = 0°$ to 100% at $\theta = 90°$. Tracking this path can yield a reasonable estimate of the average drag factor or the effective coefficient of friction. For example, a rotation of 90° to final rest would have an average attack angle of 45°. Since the $\tan 45° = .71$, an estimate of $f_e = .70 ft$ to $0.75 f$ would be a reasonable estimate. The effect of rotational losses of angular energy has been ignored.

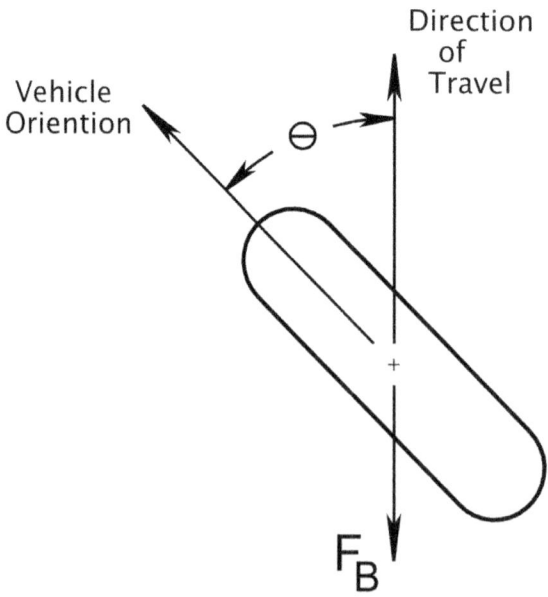

Figure 3.6. Slow Yaw

An example of a passenger car in a counterclockwise yaw is shown in Figure 3.7 and a similar yaw by a tractor-trailer is shown in Figure 3.8. (Note the tearing of grass in the median.)

Figure 3.7. Slow Yaw – Passenger Car

Figure 3.8. Slow Yaw – Tractor Trailer

Post-Impact Spinout

In many cases the post-impact trajectory consists of both translation and rotation, most commonly referred to as the post-impact spinout. This vehicle movement, although unintended, is a frequent mode of deceleration.

The kinetic energy immediately after impact has both translational and rotational components. This total kinetic energy KE_T is expressed as

$$KE_T = \frac{1}{2}mV^2 + \frac{1}{2}I\omega^2$$

where

m = vehicle mass (W/g)
V = post-impact velocity
I = moment of inertia about the vertical z axis
ω = rotational velocity about the z axis.

Both the translational and rotational kinetic energy are dissipated through friction at the tire/pavement interface. During this deceleration the vehicle travels a distance d and rotates through some angle θ.

At any given point in the post-impact trajectory, these friction forces as shown in Figure 3.9 will vary with the orientation θ_1 of the vehicle relative to the velocity vector V, the steer angle θ_2 and the effect of any wheel that is jammed.

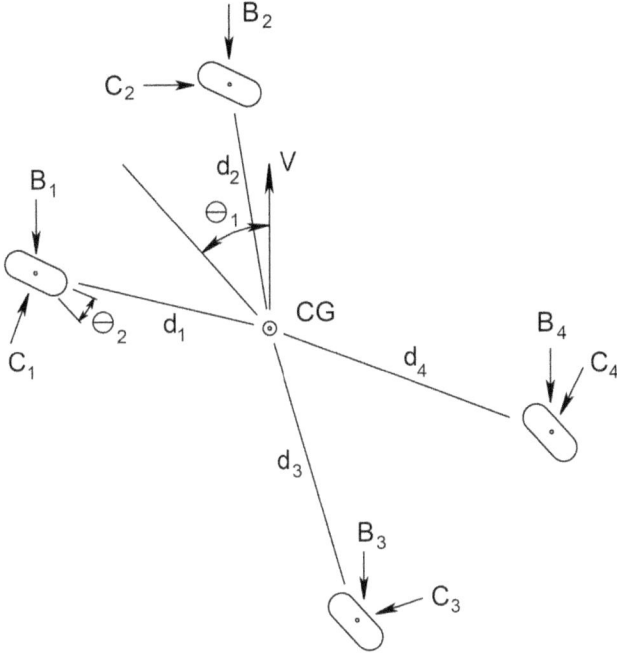

Figure 3.9. Spinout

Each of the braking forces $(B_1, B_2, B_3,$ and $B_4)$ and the rotational forces $(C_1, C_2, C_3,$ and $C_4)$ are dependent upon the coefficient of friction f and the wheel loads $(W_1, W_2, W_3,$ and $W_4)$. The magnitude of each force acting upon a free-rolling wheel is also dependent upon the orientation θ_1 of the vehicle with respect to the velocity vector V. For the front wheels the tire orientation would be equal to $\theta_1 \pm \theta_2$. The forces on any tire that is locked up and skidding will only be dependent upon the wheel loading.

Assuming that the CG lies along the longitudinal y axis, $W_1 = W_2$ and $W_3 = W_4$ when the vehicle is at rest. The static front to rear load distribution can, if unknown, be determined as previously described in **Partial Braking**. In this analysis, all forces $(B_1, B_2, B_3,$ and $B_4)$ and $(C_1, C_2, C_3,$ and $C_4)$ must be considered since each force may produce a pitch or roll moment.

Due to tire distortion and suspension flexibility, the lateral, longitudinal and vertical location of the CG will be moved from its static position. Inclusion of this change in location would require input of empirical data that may be of dubious reliability. This shift in the CG position will be particularly significant for vehicles with a high center of gravity. It should be noted that this shift in the CG location will not only affect the wheel loadings, but also will alter the moment arms of forces producing torque about the changed vertical z axis.

Throughout the post-impact spinout, rotation and curvilinear translation are occurring simultaneously, thus causing a continuous change in the relative orientation of the vehicle with respect to the velocity vector V and the position of the center of gravity. This will produce a continuous change in the wheel loadings and the magnitude, direction and moment arm of each force.

For analysis purposes it is convenient to consider the forces due to translation and the forces due to rotation independently. At any point in the trajectory the summation of the moments of all forces about the z axis will produce a torque T opposing the rotation. For a small angular change $\Delta\theta$ the torque may be considered as a constant. The work done during this small rotation will be equal to

$$\Delta\text{Work} = T\Delta\theta.$$

Determination of the total work done during the total rotation θ could possibly then lead us to a solution yielding the initial rate of rotation ω_0 and the rotation duration time t. The derivation of a general equation does not appear to be feasible given the complex relationships between the previously described variables.

One possible solution might be achieved by using numerical integration to calculate the summation of the work done due to rotation. That is

$$\text{Total work (Rotation)} = \sum_0^\theta T\Delta\theta.$$

Since the torque T is a function of the rotation and the direction of the velocity vector V, some iterative procedure *might* make it possible to correlate the time elapsed and the distance traveled due to the rotational and the translational effects. This would then allow for a solution.

This total work done due to rotation is also equal to the initial rotational kinetic energy

$$KE = \tfrac{1}{2} I_z \omega_0^2 .$$

Where I_z is the moment of inertia of the vehicle about the Z axis and is defined by

$$I_z = \int r^2 dm$$

where

dm is any element of mass

r is the perpendicular distance from the axis to the element in question.

That is

$$\tfrac{1}{2} I_z \omega_0^2 = \sum_0^\theta T \Delta \theta$$

or

$$\omega_o = \sqrt{\frac{2}{I_z} \sum_0^\theta T \Delta \theta} .$$

For a small change in the angular velocity $\Delta \omega$ the corresponding time elapsed Δt is given by

$$\Delta t = \Delta \omega / \alpha$$

where

$\alpha =$ angular acceleration rate.

From *Newton's Law*, the angular acceleration is

$$\alpha = T / I_z$$

and

$$\Delta t = \frac{I_z}{T} \Delta \omega .$$

The total time t for the total rotation θ is then given by

$$t = \sum_0^{\omega_0} \frac{I_z}{T} \Delta\omega .$$

It is recognized that achieving a solution for the rotational kinetic energy and the time of rotation might not be feasible.

A similar solution *might* also be possible for determining the initial velocity V_0 and the duration of the translation. At any point along the vehicle trajectory the force F_V opposing the velocity vector V would be the summation of the relevant components of all the friction forces. This force may be considered as constant acting over some small distance. Δd . The work done due to this small translation would then be equal to

$$\Delta\text{work} = F_V \Delta d .$$

The total work done due to translation over the total distance might also be found by numerical integration and an iterative procedure correlating the effect of time elapsed and the distance traveled upon the direction of the velocity vector V and the rotation angle θ that determine the magnitude of the of the braking force F_V .

Equating the total work done to the total loss in translational kinetic energy

$$KE = \frac{W}{2g} V_0^2$$

yields

$$\frac{W}{2g} V_0^2 = \sum_0^d F_V \Delta d$$

or

$$V_0 = \sqrt{\frac{2g}{W} \sum_0^d F_V \Delta d}$$

where

V_0 = post-impact departure speed.

During a small change in translation velocity ΔV, the corresponding time elapsed Δt is given by

$$\Delta t = \Delta V / a$$

where

    a = acceleration rate

From *Newton's Law*, the force F_V is equal to:

$$F_V = Ma = \frac{W}{g} a$$

or

$$a = \frac{g}{W} F_V$$

where

    W = vehicle weight
    F_V = braking force.

Substituting yields

$$\Delta t = \frac{W}{gF_V} \Delta V.$$

The total duration of the deceleration would be equal to

$$t = \frac{W}{g} \sum_0^{V_o} \frac{\Delta V}{V}$$

where

    V_o is the initial post-impact velocity.

Again it is recognized that a solution for t might also not be feasible.

Another complication for this possible solution is that the time duration of the rotation is not necessarily equal to the translational deceleration time. The previously described iterative procedure would have to include a correlation of the timing between the translation and rotation.

Any possible solution would obviously be so cumbersome that a hand calculation would be totally unfeasible. Computer assistance would be required but would not guarantee a successful solution. I realize that, due to the complexities involved, the preceding attempt to provide an analysis of the spinout dynamics may well be an exercise in futility, but hopefully researchers can develop a useful solution for this complex vehicle motion.

Over the years some simplified theoretical solutions have been developed to predict spinout trajectories and to estimate departure speeds. These solutions generally involved assumptions such as the following

- The CG lies at midpoint of the longitudinal axis;
- The vehicle is a rigid body;
- Lateral load redistribution can be ignored;
- Longitudinal load redistribution can be ignored;
- All wheels are free rolling;
- Steer angle is ignored.

These simplified analyses were still sufficiently complex to require a computer- assisted solution.

There has been some model testing as well as some full scale testing of spinouts. Due to these simplifying assumptions there has not, as expected, been complete agreement with the theoretical solutions. This observation is in no way intended as a criticism since these theoretical and empirical endeavors have added a great deal to our understanding of the spinout. I certainly have no definitive suggestions to improve upon these efforts.

For reconstruction purposes we are interested in calculating a reasonable estimate of the post-impact departure speed. One practical approach is to treat the spinout in the same fashion as partial braking, using an estimated average or effective coefficient of friction f_e and deceleration distance d. The post-impact speed would then be given by the expression $S = 5.47\sqrt{f_e d}$ that was derived previously for a simple locked wheel skid to a stop.

The distance d should not include any post-rotation roll to final rest. This roll portion may be considered as a second mechanism of energy loss or simply ignored since the actual kinetic loss is generally quite small.

The collective theoretical and empirical efforts indicate that an effective drag coefficient of $f_e = 0.70$ to $0.75 f$ for a free-wheeling vehicle is a reasonable value. With the jamming of one or more tires and post-impact braking this value could increase upward to $f_e = f$ for full braking. If the spinout occurs on a grade the use of $f'_e = f_e \pm g$ is a reasonable solution. Obviously a spinout analysis is not required for the post-impact or the loss of control trajectory of a motorcycle. It will simply be sliding on its side with a difficult to determine coefficient of friction of approximately 0.30 to 0.50.

Rollover

Vehicle rollovers are another unintended method of deceleration. Vehicles with relatively high centers of gravity can rollover due to traveling too fast on a curve or from the tipping encountered in a spinout. Lower profile vehicles will tend to spinout rather than rollover in these situations. A rollover is generally a less desirable outcome due to the increased potential for damage and serious injuries.

A low profile vehicle can, however, roll over due to a collision with a rigid object or another vehicle or due to some tripping mechanism such as striking a curb or furrowing into soft ground.

Initiation of the Rollover

Lifting of the inside tires on a curve or the trailing wheels in a spinout can be defined as the initiation of tipping. For a rigid body vehicle, as shown in Figure 3.10, this occurs when the overturning couple defined as the inertial force F_I times the height h of the center of gravity exceeds the restoring couple defined as the weight of the vehicle times the distance between the center of gravity and the outside of the tire.

76 | HIGHWAY ACCIDENTS

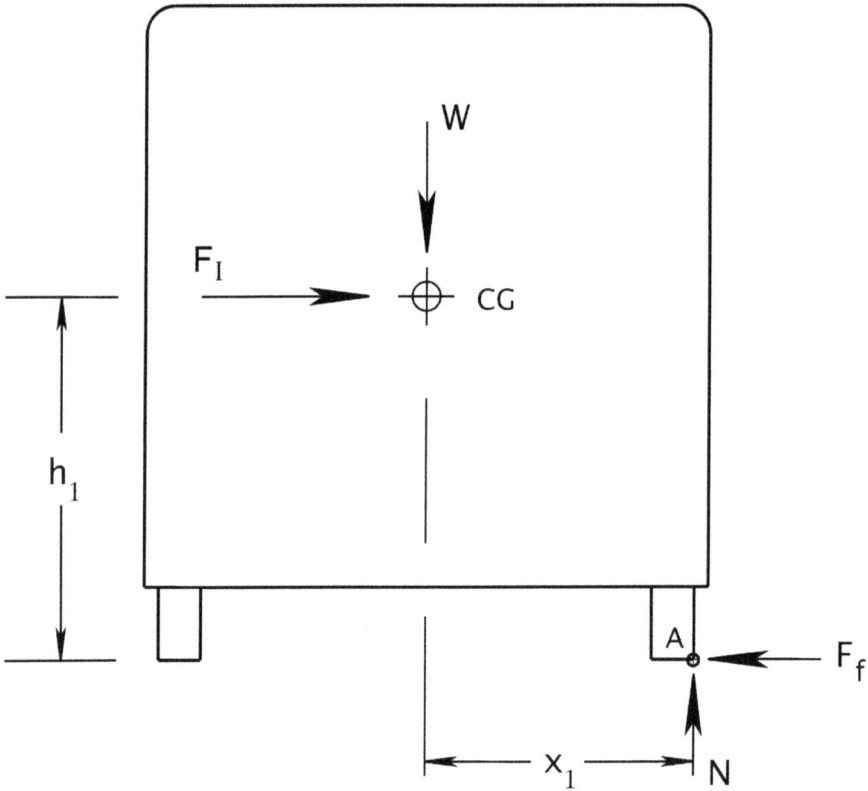

Figure 3.10. Impending Tipping

For equilibrium at impending tipping, summing forces in the x direction yields

$$\sum F_X = F_I - F_f = 0$$

or

$$F_I = F_f$$

where

F_I = inertial force
F_f = friction force = fW.

Summing forces in the y direction yields:

$$\sum F_y = N - W = 0$$

or

$$N = W$$

that is, all the weight will be on the outside tire.

Summing moments about point A yields:

$$\sum M_A = W[x_1] - fW[h_1]$$

or

$$f = x_1 / h_1 \qquad (3.15)$$

where

f = friction demand for impending tipping.

Since no vehicle is a rigid body, the actual friction demand required to imitate a rollover will always be less than that given by Equation (3.15). The inertial force will cause a small lateral shift due to bending of the tires and distortion of the suspension system. The overturning moment will produce a compression and expansion of the tires and springs, resulting in a rotation of the vehicle and, therefore, a further lateral displacement of the CG. In some situations there might be a shift of the cargo (solid or liquid) that also produces an additional lateral displacement.

In the case of a tractor-trailer combination, slack at the fifth wheel and tractor rotation will also contribute to rotation and, therefore, the translation of the center of gravity. In some vehicles, the design may permit the CG to rise somewhat during the rotation.

Due to some or all of these factors, the horizontal distance x_1 will be reduced to x_1' and the height of the CG might be increased to h_1'. This will result in the actual friction demand f' being equal to

$$f' = x_1' / h_1'.$$

This reduction may be some relatively small value for low profile automobiles to as much as 50% for some low stability vehicles, such as large cargo trucks.

If this friction demand f' is exceeded, the vehicle will start to rotate about the outside or leading tire (point A). Since this rotation will continue to decrease x_1' and increase h_1', the rollover will be completed.

Tripping

Although a low profile vehicle will normally remain upright in a side slide, a rollover may be induced by some tripping mechanism such as a curb, some other object or furrowing into soft ground. A minimum lateral speed is required to produce this tripping action.

The actual rollover will occur when the vehicle's center of gravity passes through its maximum height, thus committing the vehicle to continue the rotation. Referring to Figure 3.11, the relevant dimensions are

h_1 = height of the CG at first contact
y_1 = height of the CG above the tipping point at first contact
y_2 = maximum height of the CG above the tipping point
x_1 = horizontal distance from the CG to the external contact point A.

VEHICLE DYNAMICS | 79

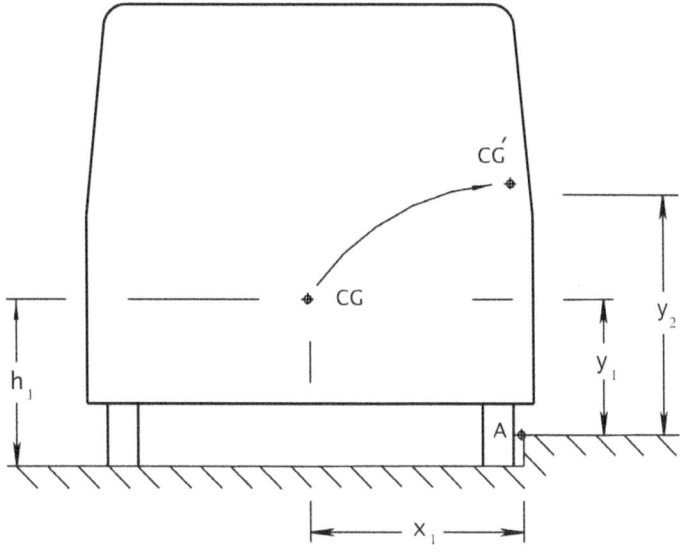

Figure 3.11. Tripping

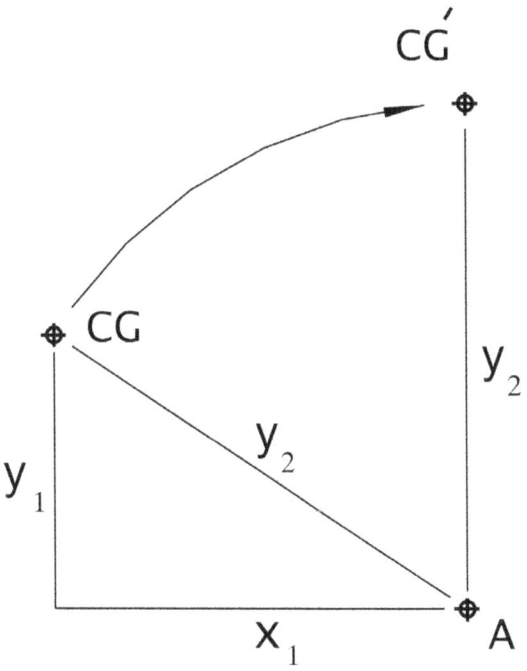

Figure 3.12. Geometry of Tripping

From the geometry shown in Figure 3.12

$$y_2 = \sqrt{x_1^2 + y_1^2}.$$

Using conservation of energy, the minimum kinetic energy required to raise the CG to its maximum height y_2 is given by

$$KE = \frac{W}{2g}V^2 \qquad (3.16)$$

where V is the minimum speed to complete the rollover.

The corresponding increase in potential energy PE is equal to

$$\Delta PE = W(y_2 - y_1). \qquad (3.17)$$

Equating (3.16) and (3.17) yields

$$\frac{W}{2g}V^2 = W(y_2 - y_1).$$

This reduces to

$$V = \sqrt{2g(y_2 - y_1)}$$

or

$$V = 8\sqrt{y_2 - y_1}$$

where

$$y_2 = \sqrt{x_1^2 + y_1^2}$$

and

V = minimum speed to produce a rollover.

For example, if $x_1 = 3.0$ ft and $y_1 = 1.0$ ft, then

$$y_2 = \sqrt{(3)^2 + (1)^2} = 3.16 \text{ ft}$$
$$V = 8\sqrt{3.16 - 1.00}$$
$$V = 11.8 \text{ ft/sec}$$

or

$$S = 8.0 \text{ mph}$$

where S is the theoretical minimum speed required to trip the vehicle and produce a rollover.

It is possible that bending of the contacting tire(s) or wheel(s) and/or distortion of the suspension may decrease the value of x_1 and, therefore, reduce the speed required to complete the rollover. If the speed at impact with the tripping mechanism is less than that required to produce a complete rollover, the trailing tires might still lift off the ground and then fall back to final rest.

Depending upon the speed and the precipitating cause of the rollover, several different movements may occur. The initial motion may be a flip or vault for some distance without any ground contact or loss of kinetic energy. It is important to realize that a roll over the right side may result in the first contact and crush to the roof on the left side of the vehicle, leaving the right side untouched. (See Figure 3.13.)

Figure 3.13. Rollover – First Contact

Motion during the rollover may include yawing about the z axis, pitching about the x axis, rolling about the y axis and translation of the center of gravity. There will also be some sliding (often called slippage) or rotating at the ground contact by any part of the vehicle. Other frequent occurrences are that sliding on the roof or a side or free rolling upright will continue after the conclusion of the rollover.

The complexity of the rollover is obviously much greater than the spinout. In many situations it is not a simple task to perform even a qualitative reconstruction. The correlation of vehicle damage such as crush and scrape marks to ground contact evidence is often quite difficult. In a high-speed rollover determination of the number of rolls may not even be obvious.

A complete quantitative description of the kinematics involved is generally not possible. Therefore, a determination of the total dynamics of the accident is not realistic. A reasonable goal is to achieve as complete as possible a qualitative description of the rollover event. Some computer simulations are excellent and provide important insight into the vehicle behavior during a rollover but, as yet, are not sufficient for an accurate quantitative reconstruction of the event.

Since the rate of loss of kinetic energy due to spinning and sliding and due to damage at contact points varies throughout the rollover, a direct analysis is most probably not feasible. It is convenient, and probably necessary, to treat the rollover in the same manner as the spinout. A reasonable estimate of an average or overall drag coefficient f_e will permit a simple calculation of the speed at the beginning of the rollover. That is

$$S = 5.47\sqrt{f_e d}$$

where

d = distance traveled from lift off to termination of the rollover.

This speed or loss in kinetic energy may be combined with losses due to any subsequent roll out or slide and/or any pre-rollover partial spinout by using the combined speed concept.

Testing has shown that an effective coefficient of friction f_e for most rollovers will range from approximately 0.40 to 0.60. If the rollover occurs on a slope this can be accounted for by using $f_e' = f_e \pm g$.

Impact Dynamics 4

Prior to proceeding with the reconstruction analysis it is also necessary to develop an understanding of the dynamics of the second phase of the accident event. This impact or collision phase is defined as the period from first contact to separation or when the impact forces go to zero. There can be large energy losses and significant momentum transfers between the vehicles with, unlike the pre-crash and post-crash phases, little interaction with the ground. Although the duration of the impact phase is generally quite short (approximately 0.10 seconds), there are often huge forces and accelerations, as well as significant changes in speed, direction, and rotation.

There are basically two methods of analysis for reconstructing the impact phase. One method is the application of the concept of conservation of momentum for vehicle-to-vehicle impacts. The other method is the utilization of the work-energy relationship that can be applied to a single-vehicle impact or a two-vehicle collision. We will first consider the work-energy approach for the utilization of damage in the single-vehicle crash.

Fixed Barrier Impact

Theoretically, if we know the deformation characteristics of a vehicle, we can calculate the loss in kinetic energy and, therefore, the impact speed when the vehicle strikes a rigid object. A major difficulty is that an automobile is a totally non-homogeneous complex structure composed of a multitude of dissimilar components. Depending upon the vehicle orientation and the size and shape of the rigid object, these components may be subjected to tension, compression, bending or torsion that can result in a change in size or shape of the overall structure.

Another problem arises since the vehicle does not respond as a purely elastic or a purely plastic material. A very low speed impact such as bumping into a pole while parking will produce a force on the vehicle and a corresponding deceleration but may cause no permanent deformation. This is an elastic impact and provides no useful information except that the speed was below that value required to exceed the elastic limit.

In a severe impact, such as driving head-on into a tree or wall at highway speed, there will be the initial small elastic deformation plus a large permanent plastic deformation. This permanent crush can be used to determine a reasonable estimate of the kinetic energy loss and, therefore, the speed at impact.

Since theoretical solutions are not feasible, extensive testing has been conducted for many years on a wide variety of automobiles for a few simple impact configurations. Testing of new vehicles for compliance with safety standards involves a head-on impact into a rigid wall. Similar tests have been conducted for research purposes with a variety of objectives. In all of these tests the vehicle weight W, the impact speed S, and the resulting crush c are recorded. These test results can be used directly if the accident vehicle damage is approximately the same as a similar test vehicle.

Plotting the results of these tests for similar vehicles at various speeds will provide data points such as shown in Figure 4.1. The dashed line is an assumed "best fit" for the test data, thus indicating an approximately linear relationship between the impact speed and the resulting crush.

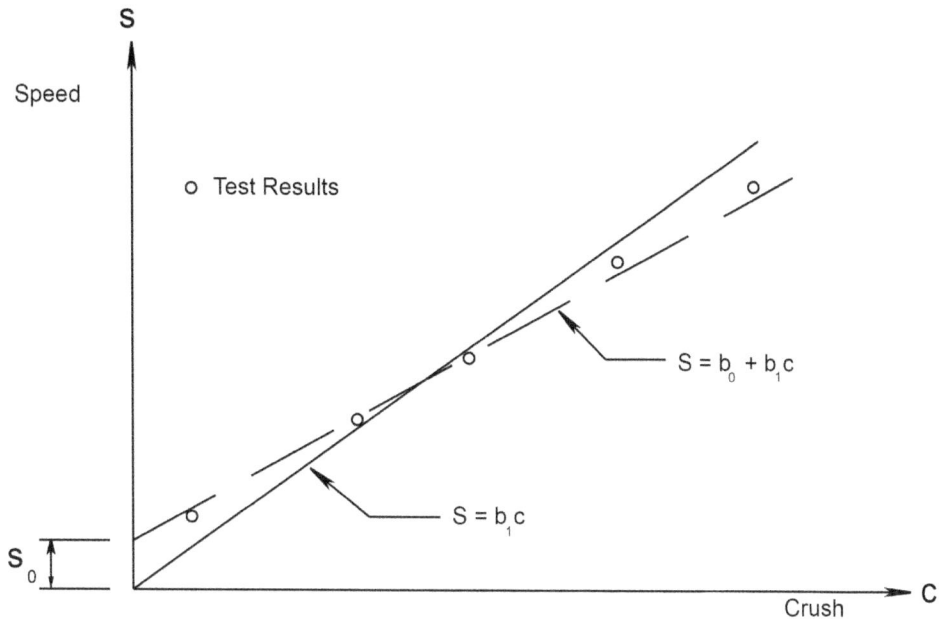

Figure 4.1. Speed vs. Crush

The slope of the line (dS/dc) is essentially the linear plastic spring constant for the inelastic range. S_0 is the speed required to reach the elastic limit and begin to produce permanent deformation. The equation of this dashed straight line is of the form

$$S = b_0 + b_1 c$$

where

S = impact speed in mph
c = crush in inches
b_0 = S_0 in mph
b_1 = slope of the line dS/dc in mph/in.

This impact speed S is also referred to as the Fixed Barrier Equivalent (FBE) speed. This concept will be used in later discussions.

For high-speed impacts it is convenient and not unreasonable to ignore the elastic deformation and assume that $S_0 = 0$. The equation of this "best fit" line would then reduce to the form

$$S = b_1 c$$

This expression of the relationship of speed versus crush is shown as the solid line in Figure 4.1. Note that the slope b_1 of this line will be slightly greater than the slope of the dashed line, and will underestimate the speed of low speed impacts and possibly overestimate the speed of high speed impacts, but will certainly give reasonable values for mid-range speeds.

This simplification allows us to consider the vehicle as a linear plastic spring. Recall that the force F required to produce a given deformation is defined as

$$F = kx$$

where

F = force in lb
x = deformation in inches
k = spring constant in lb/in.

The work done in causing a deformation c is

$$\text{Work} = \int_0^c F dx$$

or

$$\text{Work} = \int_0^c kx \, dx .$$

Integrating yields

$$\text{Work} = \frac{1}{2} kx^2 \Big|_0^c$$

or

$$\text{Work} = \frac{1}{2} kc^2 . \qquad (4.1)$$

During the deceleration there will be a loss in kinetic energy KE equal to

$$KE = \frac{W}{2g} V^2 \qquad (4.2)$$

where V is the impact speed in ft/sec.

Since the work done is equal to the loss in kinetic energy, we can equate (4.1) and (4.2), thus yielding

$$\frac{W}{2g} V^2 = \frac{k}{2} c^2$$

or

$$V^2 = \frac{gk}{W} c^2$$

or

$$V = c \sqrt{\frac{gk}{W}} \quad \text{in ft/sec.}$$

Converting to speed S in mph yields

$$S = \frac{c}{1.47} \sqrt{\frac{gk}{W}}.$$

Since the speed is also $S = b_1 c$ or $b_1 = S/c$

$$b_1 = \frac{1}{1.47} \sqrt{\frac{gk}{W}}$$

or

$$b_1 = \frac{\sqrt{32.2}}{1.47} \sqrt{\frac{k}{W}}$$

or

$$b_1 = 3.87 \sqrt{k/W}. \qquad (4.3)$$

Later we will be using S^2 as representing the kinetic energy where

$$S^2 = b_1^2 c^2$$

or

$$S^2 = 15 \frac{k}{W} c^2$$

In using the published test results it is important that the test vehicle be similar to the accident vehicle. If the weight (including occupants and cargo) of the accident vehicle W_A is different from the test vehicle W_T, some adjustments may be required.

The test will yield an impact speed S, a crush c, and a weight W. For this weight the stiffness coefficient b_1 (test) is given by

$$b_1(test) = S/c.$$

Since b_1 is inversely proportional to the square root of the weight, as shown in Equation (4.3), the adjusted value of b_1 will be equal to

$$b_1 = b_1(test)\sqrt{W_T/W_A}$$

where

W_T = weight of the test vehicle
W_A = weight of the accident vehicle

or

$$b_1^2 = b_1^2(test) W_T / W_A.$$

The preceding discussion has been limited to a full frontal impact with a uniform crush profile and the stiffness coefficient b_1 is given by

$$b_1 = S/c.$$

For example, if the test speed was 36 mph and the crush was 24 inches, b_1 would be equal to

$$b_1 = \frac{36 mph}{24 in} = 1.5 mph/in$$

A fixed barrier crash of a similar vehicle of the same weight that produced a uniform crush of 12 inches would be due to an impact speed equal to

$$S = (1.5 \text{ mph/in})(12 \text{ inches})$$

$$S = 18 \text{ mph}.$$

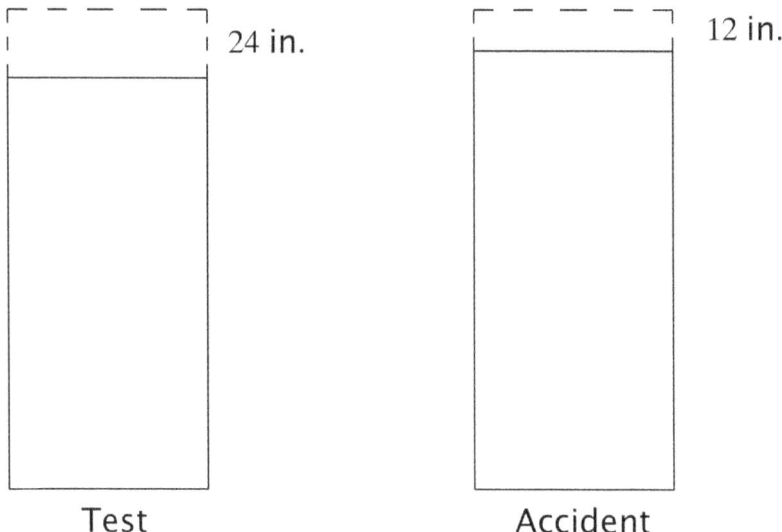

Figure 4.2. Test Profile vs. Accident Profile

Referring to Figure 4.2, the total deformed volume of the accident vehicle is 1/2 that of the test vehicle, and the impact speed is also equal to 1/2 that of the test vehicle. Since kinetic energy is proportional to S^2, the energy loss of the accident vehicle is only 1/4 of that of the test vehicle. That is,

$$\frac{(18)^2}{(36)^2} = \left(\frac{1}{2}\right)^2 = \frac{1}{4}$$

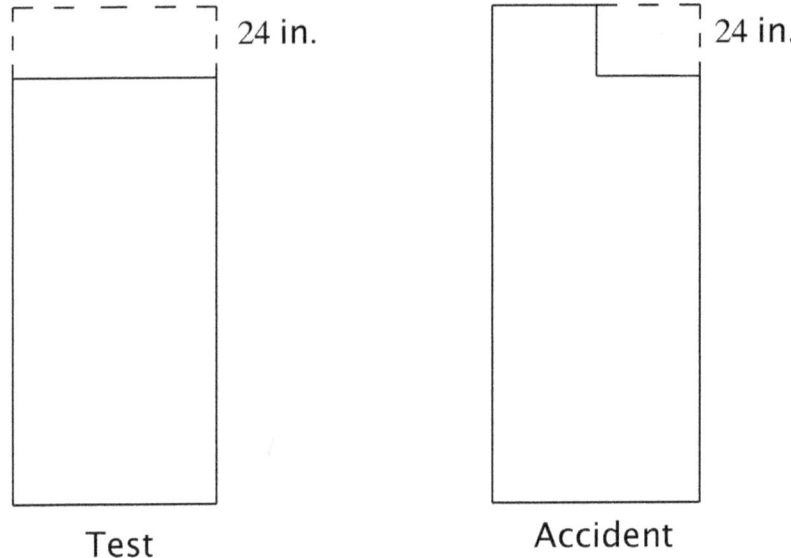

Figure 4.3. Test Profile vs. Accident Profile

In another crash with the damage pattern as shown in Figure 4.3, the deformed volume would again be equal to 1/2 of that of the test vehicle, but the kinetic energy loss would be 1/2, not 1/4, of the energy loss in Figure 4.2. The energy loss for the test vehicle is

$$KE = (36)^2 = 1296.$$

For the accident vehicle in Figure 4.3 the energy loss would therefore be equal to

$$KE = \frac{1}{2}(1296) = 648$$

and the impact speed would be

$$.S = \sqrt{648} = 25.5 \text{ mph}.$$

Note that incorrectly using the average crush of 12 inches would yield an estimated impact speed of 18 mph with a kinetic energy of

$$KE = (18)^2 = 324.$$

IMPACT DYNAMICS | 93

Using the average crush in this case would produce an unacceptable error.

It is important to remember that we are dealing with a loss in kinetic energy which is proportional to speed squared and can be expressed as

$$S^2 = b_1^2 c^2.$$

In our example for the test vehicle

$$S^2 = (1.5)^2 (24)^2 = 1296$$

or

$$S = 36 \text{ mph}$$

Another approach is to consider the stiffness coefficient b_1^2 per unit width. Assuming a vehicle width equal to 60 inches, the value for b_1^2 / in would be

$$b_1^2 \text{ / in} = (1.5)^2 / 60 = 0.0375 / \text{in}$$

For the accident vehicle in Figure 4.3 the kinetic energy is then equal to

$$S^2 = (.0375)(30)(24)^2 = 648$$

or

$$S = 25.5 \text{ mph}.$$

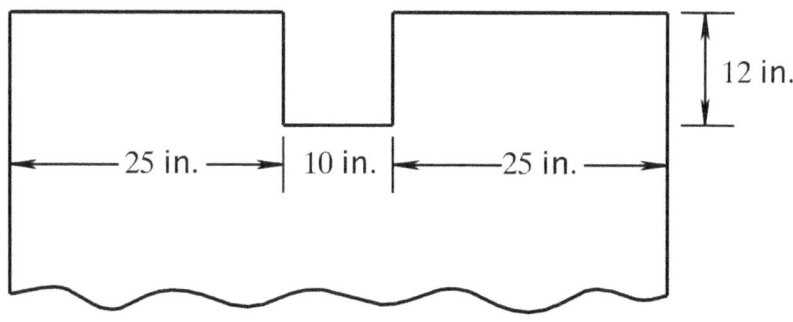

Figure 4.4. Narrow Impact

For a narrow impact, such as striking a pole, the crush profile could be shown as in Figure 4.4. In this case the kinetic energy would be equal to

$$S^2 = b_1(10)c^2$$

$$S^2 = (.0375)(10)(12)^2 = 54$$

or

$$S = 7.3 \text{ mph}.$$

It should be noted that the stiffness coefficient for narrow impacts will actually vary across the front of the vehicle, thus leading to a potential error.

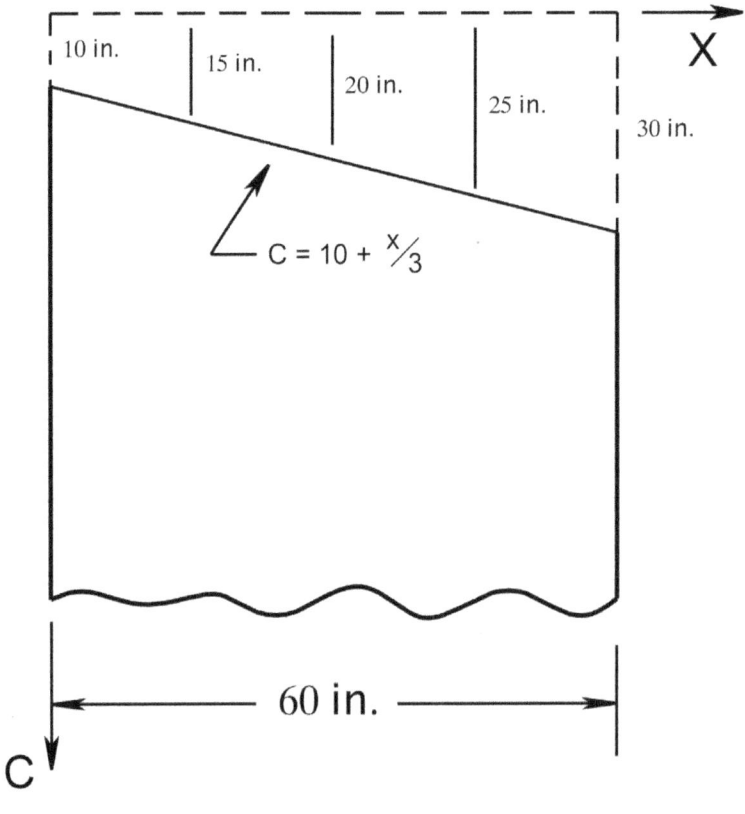

Figure 4.5. Angle Impact

Striking a barrier at an angle could produce a crush profile such as shown in Figure 4.5. Using the average crush of 20 inches would yield a kinetic energy loss equal to

$$S_1^2 = .0375(60)(20)^2 = 900$$

or an estimated speed $S = 30$ mph. Since from a previous example we realize that this will actually underestimate the actual speed, it might be appropriate to divide this into four sections as also shown in Figure 4.5. Using the average crush for each section and summing the four kinetic energies yields

$$S_1^2 = .0375(15)(27.5)^2 = 425$$

$$S_2^2 = .0375(15)(22.5)^2 = 285$$

$$S_3^2 = .0375(15)(17.5)^2 = 172$$

$$S_4^2 = .0375(15)(12.5)^2 = 88$$

$$S^2 = \sum S_n^2 = 970$$

and

$$S = 31.1 \text{ mph}.$$

Smaller sections could be utilized in this numerical integration procedure to refine the speed value with little increase in accuracy. Since the crush profile can be described by a simple equation, a direct approach is also possible. The crush is given by the following equation

$$c = 10 + x/3$$

$$c^2 = 100 + 6.67x + x^2/9.$$

The kinetic energy dS^2 for a segment of width dx is equal to

$$dS^2 = .0375\left(100 + 6.67x + x^2/9\right)dx$$

or

$$S^2 = .0375\int_0^{60}\left(100 + 6.67x + x^2/9\right)dx.$$

Integrating yields

$$S^2 = .0375\left[100x + 3.33x^2 + x^3/27\right]_0^{60}$$

$$S^2 = .0375\left[100(60) + 3.33(60) + (60)^3/27\right]$$

$$S^2 = 975$$

$$S = 31.2 \text{ mph}.$$

It is apparent that, in this case, using the average crush (20 inches) yields a speed estimate that is only off by

$$Error = (31.2 - 30.0)/31.2 = .0385$$

$$Error \approx 4\%.$$

This error may actually be less than the error in the stiffness coefficient b_1 obtained from the crash testing. There is essentially no error introduced if the numerical integration procedure utilizes four sections.

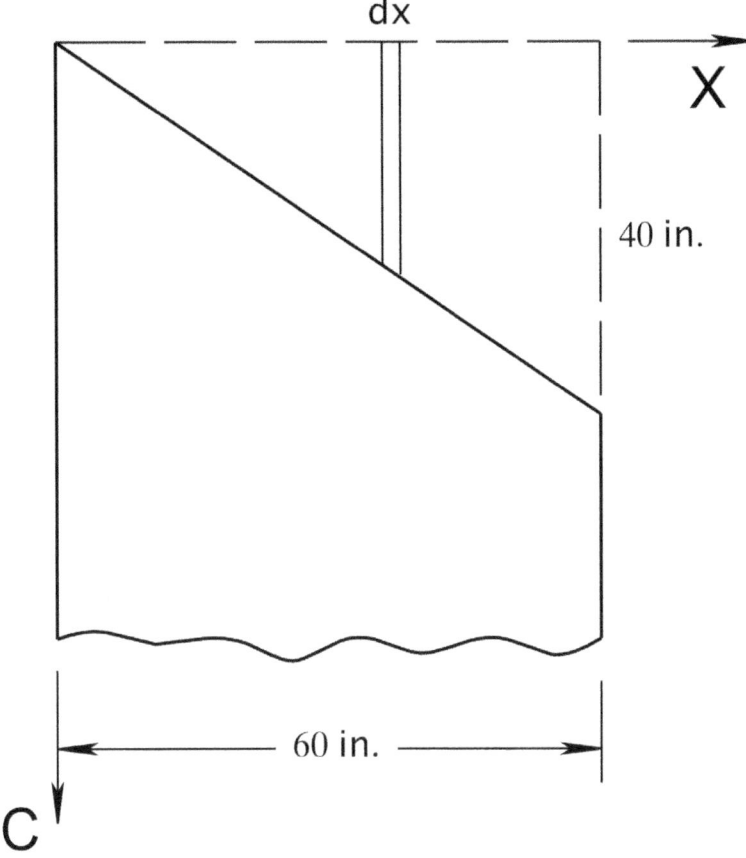

Figure 4.6. Angle Impact

A more dramatic example of a crush profile is shown in Figure 4.6. In this case using the average crush (20 inches) would again yield

$$S^2 = 900$$

$$S = 30 \text{ mph}$$

The equation describing this crush profile is:

$$c = (40/60)x = 0.667x$$

and

$$c^2 = 0.444x^2.$$

The kinetic energy dS^2 for a segment of width dx is equal to

$$dS^2 = .0375(.444x^2)dx$$

or

$$S^2 = .0375\int_0^{60}(.444x^2)dx.$$

Integrating yields

$$S^2 = .0375[0.148]_0^{60}$$

$$S^2 = .0375(0.1481)(216,000)$$

$$S^2 = 1200$$

$$S = 34.6 \text{ mph}.$$

Using the average crush in this case produces the following error

$$Error = 4.6/34.6 = .133$$

$$Error \approx 13\%.$$

These last examples indicate that some judgment needs to be exercised in selecting the methodology. For even complex damage profiles using only a few sections to calculate the estimated fixed barrier equivalent (FBE) speed should be sufficient. Use of the relationship $S = b_0 + b_1 c$ may well produce a much better estimate of the impact speed if the impact speeds are lower than or much higher than those used for the test vehicles.

For low speed impacts the assumption that the impact can be treated as a linear plastic spring is no longer valid. Use of the equation $S = b_1 c$ ignores the elastic range and assumes that S_0 is equal to zero. As shown in Figure 4.1, for a given value of crush the speed estimate from the simplified form $S = b_1 c$ will yield a much smaller speed estimate than that given by $S = b_0 + b_1 c$.

If the crush profile for a low speed is uniform, the method of analysis may be used as previously described except we will now use

$$S = b_0 + b_1 c$$

$$S^2 = b_0 + 2 b_0 b_1 c + b_1^2 c^2 .$$

Note the slope b_1 is slightly greater than b_1 for the simplified form $S = b_1 c$.

If the crush is not uniform, we can again use the same method of dividing the profile into sections and thereby determine an estimate of the impact speed. This computation will be somewhat cumbersome but can be done with or without using a computer. In general impacts at less than 10 mph do not cause significant injuries, do not involve speeding and, therefore, do not generally require a speed analysis.

Rear impacts can be analyzed by following the same procedures as in the previously discussed frontal impacts. Determination of the kinetic energy loss and the corresponding FBE speed is, of course, dependent upon test results. Since empirical data for rear impacts is sparse, the usefulness of these analyses is questionable.

Side impact testing generally involves a rigid barrier with a shape similar to the front of an automobile. (There is some limited testing of pole impacts.) Therefore, the full height of the vehicle is not involved. Another problem is that these test impacts are between the front and rear wheels. Impacts that include axle contact will naturally have much less crush and, therefore, cannot be related to the crash test data. This type of impact is shown in Figure 4.7.

Figure 4.7. Axle Impact

Due to lack of any meaningful test data, speed estimates from damage are not feasible for trucks, buses, and motorcycles.

Vehicle To Vehicle Impacts

Although the preceding discussion has been limited to the determination of the loss of kinetic energy and the resulting FBE speed for a single vehicle, a damage analysis can also be applied to impacts between two vehicles. Conservation of energy dictates that the total kinetic energy at separation is equal to the total kinetic at impact minus the kinetic energy lost during impact. The total kinetic energy at separation is determined from the post-impact motion of the vehicles as described in Chapter 3. Vehicle Dynamics.

This energy loss during impact is due to the damage to each vehicle plus the friction losses due to the sliding of tires throughout the impact phase. The time elapsed and the distance traveled are rather small so this friction loss is generally small compared to the damage losses.

A simple and convenient method to account for this friction loss is to include it in the post-impact movement of the vehicles. This is done by assuming that the location and orientation of each vehicle after impact is the same as at impact.

An example of an impact configuration that would be amenable to an analysis using only damage would be Vehicle #1 (V1) striking a stopped Vehicle #2 (V2) in the classic perpendicular "T-bone." Conservation of energy dictates that the total kinetic energy at impact KE_I is equal to kinetic energy lost by both vehicles due to damage during impact KE_D plus the total kinetic energy of both vehicles after impact KE_P. Since V2 is stopped, the total kinetic energy at impact KE_I would be due to V1. That is, $KE_I = W_1 S1_1^2$ where $S1_1$ is the speed of V1 at impact. The kinetic energy loss for the full frontal crush to Vehicle #1 plus the kinetic energy loss for Vehicle #2 due to the direct side crush would yield the total energy loss during impact KE_I.

Since the post-impact speeds of both vehicles would essentially be the same known value $S1_2 = S2_2 = S_2$, the post-impact kinetic energy KE_P would be given by

$$KE_P = (W_1 + W_2)S_2^2$$

where

KE_P = total post-impact kinetic energy
W_1 = weight of V1
W_2 = weight of V2
$S1_2$ = post-impact speed of V1
$S2_2$ = post-impact speed of V2
S_2 = composite post-impact speed.

The loss in kinetic energy KE_D due to damage is equal to

$$KE_D = W_1 S1_B^2 + W_2 S2_B^2$$

where

$S1_B$ = calculated fixed barrier speed of V1 as previously described

$S2_B$ = calculated fixed barrier speed of V2 as previously described.

Other examples of the use of a damage analysis will be given later.

Conservation of Momentum

A powerful tool that may be used in the analysis of two-vehicle impacts is the principle of conservation of momentum. This states that "the total momentum of both vehicles at impact is equal to the total momentum of both vehicles after impact." This can be expressed in the form of an equation

$$m_1 S1_1 + m_2 S2_1 = m_1 S1_2 + m_2 S2_2$$

where

m_1 and m_2 = masses of V1 and V2, respectively
$S1_1$ = Pre-crash speed of V1
$S2_1$ = Pre-crash speed of V2
$S1_2$ = Post-impact speed of V1
$S2_2$ = Post-impact speed of V2.

Since $m = Wg$ and g is a constant, this momentum relationship can be expressed as

$$S1_1 W_1 + S2_1 W_2 = S1_2 W_1 + S2_2 W_2.$$

Momentum is a vector quantity, therefore, it is generally convenient to express this relationship in terms of its x and y components. Although this may appear inelegant and cumbersome it is felt that using the speed components which are scalar values allows for a better feel for the analysis. It also simplifies the checking of the calculated impact speed and the corresponding changes in speed (ΔS's) during impact.

Conservation of momentum in the x direction may now be expressed as

$$S1_1(x)W_1 + S2_1(x)W_2 = S1_2(x)W_1 + S2_2(x)W_2. \quad (4.4)$$

Similarly in the y direction

$$S1_1(y)W_1 + S2_1(y)W_2 = S1_2(y)W_1 + S2_2(y)W_2 \quad (4.5)$$

where

$S1_1(x) = x$ component of the impact speed of Vehicle #1
$S1_1(y) = y$ component of the impact speed of Vehicle #1
$S2_1(x) = x$ component of the impact speed of Vehicle #2
$S2_1(y) = y$ component of the impact speed of Vehicle #2
$S1_2(x) = x$ component of the departure speed of Vehicle #1
$S1_2(y) = y$ component of the departure speed of Vehicle #1
$S2_2(x) = x$ component of the departure speed of Vehicle #2
$S2_2(y) = y$ component of the departure speed of Vehicle #2
W_1 = weight of Vehicle #1
W_2 = weight of Vehicle #2.

The objective is to find the four impact speeds $S1_1(x)$, $S1_1(y)$, $S2_1(x)$ and $S2_1(y)$. We now have two independent equations (4.4) and (4.5) and four unknown impact speeds. In order to achieve a solution one of the following situations is required:

1. Two of the approach speed components are known, thus reducing the solution to two equations in two unknowns;
2. The approach angles θ_1 and θ_2 of both vehicles relative to the x axis of our reference system are known since

$$S1_1(y) = S1_1(x) \tan \theta_1 \quad (4.6)$$

$$S2_1(y) = S2_1(x) \tan \theta_2. \quad (4.7)$$

We now have four equations in four unknowns.

3 One approach speed component of one vehicle is known and the direction of the other vehicle is known. This would then yield three equations in three unknowns.

4 One approach speed component of one vehicle is known and the damage patterns to the vehicles are sufficient to give a reasonable estimate of their *relative* approach angles. This is not only an infrequent occurrence but also requires some input from the damage pattern.

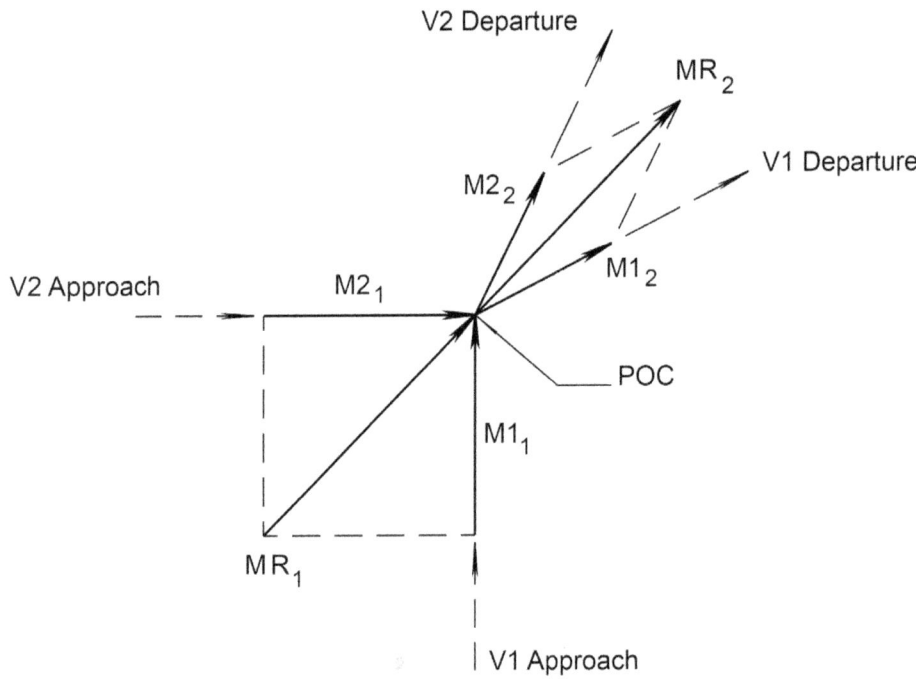

Figure 4.8. Graphical Solution

Since momentum is a vector quantity with magnitude and direction, a graphical solution, with or without computer assistance, is another option. This can also be used as a simple visual check on the previous calculations. For example, in situation 2) where the magnitude and direction of the post-impact speed and, therefore, the post-impact momentum of each vehicle, as well as their pre-crash approach directions are known. The graphical solution, as shown in Figure 4.8, would be as follows:

1. Plot the post-impact momentum of Vehicle #1 and Vehicle #2 where

$$M1_2 = S1_2 W_1 \text{ and } M2_2 = S2_2 W_2.$$

2. Using the vector parallelogram, find the resultant total post impact momentum MR_2.

3. From conservation of momentum $MR_1 = MR_2$.

4. Using the vector parallelogram as shown with the vehicle approach directions, $M1_1$ and $M2_1$, are determined.

5. The pre-crash or impact speeds of V1 and V2 are then given by

$$S1_1 = M1_1 / W_1 \text{ and } S2_1 = M2_1 / W_2.$$

Using conservation of momentum has the major advantage that the work done and the kinetic energy loss during impact are irrelevant and, therefore, the damage can be ignored. A corollary benefit is that this method applies to both elastic and plastic collisions.

The major limitation of using conservation of momentum to calculate impact speeds is that the approach and departure directions of both vehicles must be known. If a reasonable estimate of these directions cannot be obtained from the physical evidence, then conservation of momentum cannot be used.

It is important to realize that, when qualitatively analyzing a vehicle-to-vehicle collision, the resulting post impact departure directions may be substantially influenced by rotation of the vehicle which places the tires on a new path. This is not particle dynamics.

A useful concept in the qualitative analysis of an impact comes from Newton's third law of motion which can be stated as "the force that body A exerts on body B is equal to and opposite of the force that body B exerts on body A." This is shown in Figure 4.9.

(a)

(b)

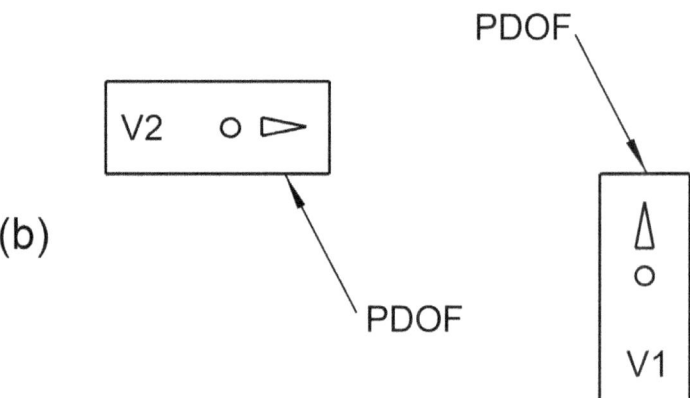

Figure 4.9. Principal Direction of Force

The resultant forces F_1 and F_2, due to the friction and normal forces between the vehicles are equal and opposite. The direction of this resultant force is the PDOF or the principal direction of force. If the PDOF for a given vehicle does not pass through the center of gravity, the vehicle will experience a change in rotation as well as a change in speed and direction.

Since the quantitative impact analysis has provided the pre-crash impact velocity, this value can be compared with the post-impact velocity to yield the change in velocity during impact. This change in velocity is generally referred to as delta V or ΔV.

In some research situations it is desirable to determine this force acting upon and/or the accelerations experienced by the accident vehicle. Since the longitudinal and lateral accelerations are not constant throughout the impact phase, and the time duration of the impact is unknown, it is not possible to calculate these values from the speed changes alone. Accelerometers are used in crash tests to determine the duration, peak value, and shape of the acceleration pulse.

Occupant Dynamics

Injuries occur during collisions when the occupant:

- Strikes the interior of the vehicle
- Is decelerated by seatbelts or airbags
- Is ejected from the vehicle.

In order to understand the injury causation mechanisms, it is necessary to have a reasonable reconstruction of the occupant kinematics. This occupant motion is a function of the lateral and longitudinal changes in speed of the vehicle that essentially determine the Principal Direction of Force (PDOF). An occupant will move in the opposite direction to the PDOF acting upon the vehicle which is now our frame of reference or coordinate system for occupant kinematics.

Placing yourself, literally or virtually, in the position of the occupant allows you to visualize the motion of the body and what it might strike and then decelerate to a stop. This would then lead to an examination of the appropriate area of the vehicle interior as described in the Collection of Evidence section of Chapter 2. This would include, but not be limited to, the following:

- Windshield fractures (generally star-shaped)
- Distortions of the steering wheel, gear shift, or rear view mirror
- Damage to the dash, doors, side panels, seat backs, or roof
- Contact points indicated by smears, cloth impressions, hair, blood, or tissue

- Ejection paths such as doors, windows, or sun roof
- Evidence indicating air bag deployment, or restraint system usage.

Combining the predicted direction of motion of the occupant with the identified points of contact generally completes the description of the occupant kinematics.

After this qualitative evaluation of the occupant kinematics, the forces involved can be addressed. It is important to understand that the severity of the impact as measured by the vehicle acceleration pulse is not the same as that experienced by the occupant. In general the most important criteria for injury production are the impact configuration and the change in speed of the vehicle. It is important to remember that change in speed, not the vehicle impact speed, is the primary factor affecting the occupant dynamics.

A portion of the body will strike the restraint system or some interior surface of the vehicle at a speed (relative to the vehicle) closely related to the ΔV of the vehicle, but the deceleration is determined by how much that body part and the contact surface will compress. Restraint systems and compressible surfaces will extend the time duration of this second impact and, since $a = \Delta V / t$, reduce the deceleration rate, the corresponding force and, most probably, the severity of the injury. Specialists in biomechanics and individuals with medical training can then correlate the observed injuries with the reconstructed occupant kinematics. If possible, these injury specialists should participate in the vehicle inspection to assist them in determining the injury mechanisms.

Analysis Procedures 5

The primary objective of this chapter is to present a methodology for the reconstruction of a variety of highway accident impact configurations and situations. We will also address the role of analysts in the collection and interpretation of evidence, as well as their inevitable role in the identification of selected causation factors.

Reconstruction of an accident event begins with the collection and interpretation of physical evidence, as previously described. Some of this may be provided by an accident investigator and presented in the form of reports, diagrams and photographs of the scene and vehicles. Even excellent reports and photographs do not eliminate the need for the analyst to conduct a thorough site inspection and a complete inspection of the vehicles.

An inspection of the crash site should result in a detailed diagram of the physical layout of the scene. Any remaining physical evidence needs to be identified, interpreted, photographed and located on the diagram. Scene photographs taken by an investigator should be utilized to identify and locate evidence not apparent at the site and to check measurements given in the reports.

Particular emphasis should be placed on determining the location and orientation of the vehicles at impact (POC) and final rest. Pre-crash approach directions can generally be deduced from the vehicle orientation at impact. It is also essential to identify and measure the physical evidence that indicates the post-impact departure directions. Any significant error in the determination of a departure direction can invalidate the entire reconstruction effort. It is important that the analyst conduct vehicle inspections and site visits whenever feasible. Otherwise, the reconstruction exercise may become a hypothetical solution only. Without this fieldwork important accident and injury causation factors are also likely to be overlooked.

At this time it is appropriate to initiate the all-important qualitative reconstruction of the accident event that describes what happened throughout the accident sequence of events without a determination of time, distance, and speed. The existing physical evidence and/or photographs can be used to indicate pre-crash maneuvers and post-impact trajectories such as spinouts or rollovers. An understanding of the dynamics of the crash event can lead to a search for over-looked scene evidence and can raise questions that need to be addressed in the vehicle inspections.

Upon completion of the qualitative analysis, it is appropriate to start the process of determining any environmental causes of the crash. Causal factors that need to be considered obviously vary from case to case but may include one or more of the following examples:

- Sight distance problems due to blockage by vegetation, structures or parked vehicles
- Traffic signs obscured by vegetation or parked vehicles
- Inappropriate use of pavement markings, signage and barriers where construction or other roadway activities are present
- Pavement defects such as excessive wear, poor drainage or potholes
- Pavement shoulder drop-offs
- Unstable shoulders or roadsides
- Unprotected roadside hazards
- Other suspected design defects or construction/maintenance problems that require highway engineering input

Examination of the vehicle(s) should be initiated by comparing the vehicle to any on-scene photos. This could help identify any alterations due to towing or storage. The use of forms with generic vehicle sketches and a checklist can help prevent missing important pieces of evidence. A complete damage description along with corresponding photographs is required to preserve the evidence.

Some important considerations in the conduct of the vehicle inspection(s) include the following:

- The location and configuration of the damage can assist in establishing the relative orientation of vehicles at impact.
- A careful measurement of the crush profile is required for an impact damage analysis.
- Identification of snagging or sideswiping is important since this may invalidate the use of some computer programs.
- Points of contact with the pavement, ground, barriers, trees, etc. need to be identified.
- A careful examination of the interior is important for any analysis of the "second impact."
- Identification of damage or conditions that correlate with the scene evidence is helpful in verifying the qualitative reconstruction. Inconsistencies may require another look at the scene evidence.

There is some debate as to whether the site inspection or the vehicle inspection should be completed first. My experience has been that it depends upon the individual circumstances. If you are provided an opportunity to make a timely site inspection, that should probably be done first since the scene evidence will generally deteriorate more quickly.

The next step is common to most problem solving endeavors. All the information from witness statements, reports, photographs, site inspections and vehicle examinations needs to be combined with all other vital information (vehicle weights, dimensions, etc.) in an orderly coherent manner. It is necessary to start with precisely what you do know before attempting to solve a problem.

All scene information should be placed on a scale diagram of the physical site. An aerial photograph (to scale and in color) is an even better choice for four reasons:

1. Complex intersection geometry is difficult and expensive to duplicate.
2. Features that may become important in the analysis will not be omitted.
3. In any future presentations (e.g., at trial) most people will understand a photograph much better than even an excellent engineering drawing.
4. An aerial photo can provide the basic background for a computer animation.

Having assembled all the data, the quantitative accident analysis can begin.

Multi-vehicle crashes include sideswipes, head-on and rear-end impacts and angled intersection type collisions. The two-vehicle intersection collision is a good example with which to start the discussion of reconstruction procedures. The accident event can be described as consisting of three distinct phases, pre-crash, impact and post-impact. The pre-crash phase consists of any vehicle maneuvers such as accelerating, braking or swerving just prior to first contact. The impact phase is defined as lasting from first contact to separation when the forces go to zero. This phase generally has a time span of 75 to 150 milliseconds or roughly 0.1 second. During this phase there is an exchange of momentum between the vehicles and a loss of kinetic energy due to vehicle damage. The final post-impact phase begins at separation and ends at final rest. This deceleration movement may be a rollout, a spinout, a rollover or some combination that dissipates the remainder of the kinetic energy. Good problem solving practice dictates that we start from a set of known values, namely final rest, and work backwards through the accident sequence.

Post-Impact

The position and orientation of each vehicle must be determined at separation and final rest. As previously discussed, a significant error will rarely be introduced by using the impact positions as equal to the locations of the vehicles at separation.

The objective is to determine the departure speed and the departure direction of each vehicle. The direction must be determined from the physical evidence at the scene such as gouge marks, scratches, yaw marks, and fluid trails. It is essential that the determination of a reasonably accurate departure direction be given the highest priority. Placing a scaled vehicle on the diagram is helpful. This scale vehicle would include tire positions and the location of the center of gravity of the damaged vehicle. (See Figure 2.2.) This may also be done with a Computer-Assisted Design (CAD) program. It is important to remember that the direction of a given gouge mark or tire mark is not necessarily the departure direction of the vehicle's center of gravity. Since the post-impact trajectory often follows a curved path, we must remember that the departure direction is *not given by the straight line from impact to final rest*.

The departure speed is determined by the loss of kinetic energy in the post-impact rollout, spinout, rollover or some combination of these. These modes of deceleration have been discussed at some length in the Chapter 3 Vehicle Dynamics.

Having obtained the departure speed and direction, it is often convenient to separate the speeds into their x and y components, thus obtaining four known values: $S1_2(x)$, $S1_2(y)$, $S2_2(x)$, and $S2_2(y)$ where, for example, $S1_2(x)$ is the departure speed for Vehicle #1 in the x-direction.

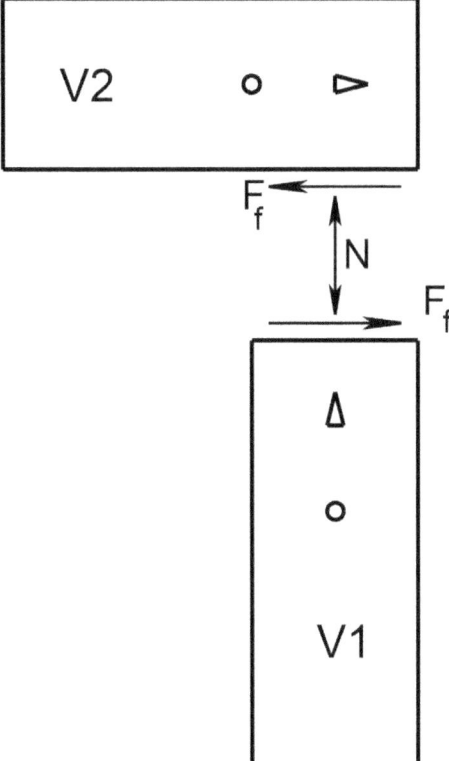

Figure 5.1. Speed Correlation

The final step in the post-impact phase of the analysis is the essential check on the correlation of the departure speeds. Referring to Figure 5.1, the equal and opposite forces acting on the two vehicles are composed of normal and friction forces. For this impact configuration $S1_2(y)$, the y component of the exit speed of V1, must be equal to or slightly less than $S2_2(y)$, the y component of the exit speed of V2. Since most impacts are plastic in nature the values of $S1_2(y)$ and $S2_2(y)$ will be close to the same value. These required conditions may be expressed by the following equations.

$$S1_2(y) \leq S2_2(y)$$

but

$$S1_2(y) \approx S2_2(y).$$

Another requirement is that due to the rotation of V1, the x component of the exit speed of V1 must be less than the x component of the exit speed of V2. That is

$$S1_2(x) < S2_2(x).$$

If these correlation requirements are not met, then obviously a computation error has been made or the assumptions or estimates used in the calculation of one or both of these speeds are not valid. It is then required to review these analyses and make the appropriate adjustments. This necessary speed correlation may be considered as a post-impact analysis tool.

This post-impact analysis may be considered as being composed of the following steps.

1. From the physical evidence we can determine for each vehicle:
   a. The location and orientation at impact (We will assume this is the same as post-impact.)
   b. The location and orientation at final rest
   c. The post-impact departure direction
   d. The post-impact departure speed
2. Using the available evidence consider the following possibilities.
   a. The driver may be braking.
   b. Partial braking may occur due to tire jamming, air out, or tread separation.

c. The driver may be steering.

d. There may be an initial steer angle.

e. The vehicle may simply roll to final rest.

f. The vehicle is in a yaw or spinout.

g. A rollover occurs.

h. Some combination of the above occurs.

From these considerations develop a complete description of the post-impact motion of each vehicle. This qualitative analysis must be correct or the entire reconstruction effort will be worthless.

3 Make a preliminary estimate of the departure speeds of each vehicle, using a best estimate of the appropriate overall effective drag factor.

4 If possible, use more sophisticated methods to analyze the various motions and combine these energy losses to compute a refined estimate for each departure speed.

5 Compare the two post-impact speeds to determine if the speed correlation criteria are met and adjust if necessary.

Impact

Having determined a reasonable estimate of the departure speed and direction of each vehicle, it is now possible to compute the speeds at impact. There are, as described in Chapter 4 Impact Dynamics, two methods for calculating the impact speeds and the corresponding changes in speed or ΔV's. Conservation of momentum is the preferred method since it is simpler and does not require a consideration of damage. The use of this method is, however, not possible if a reasonable estimate of the required approach and departure directions cannot be determined from the physical evidence.

Using the work-energy method, which is based on vehicle damage, to calculate impact speeds and/or changes in speed during the impact is more complicated and relies upon limited data from staged collisions. If the damage configurations of the accident vehicles are similar to those in the staged crashes, then this method may be used with some confidence.

Theoretical analyses and extensive testing has been conducted in an effort to use this damage analysis for two-vehicle collisions. An early example that I was involved in was the field testing of NHTSA's Crash Program during the National Crash Severity Study carried out during the late 1970's at several locations around the country.

This four-year study involved thousands of two-vehicle accidents. The results were useful for many comparison purposes but the damage-only analysis did not prove to be a reliable method for the reconstruction of many configurations of two-vehicle collisions. One important use of the damage-only analysis is to compare the resultant changes in speed (ΔV) values to the ΔV's calculated using a conservation of momentum analysis. This provides an important cross check for gross errors and should be used wherever possible.

Pre-Crash

Having determined the speed and direction of each vehicle at impact, it is now possible to analyze some pre-crash maneuvers.

Avoidance Steering

This maneuver almost always involves a sudden steering input to turn away from the intersecting vehicle. This will generally be the response even when steering in the opposite direction could have avoided the collision. This is often *followed* by a brake application since the steering input can be applied more quickly than braking.

This maneuver is not always easy to detect. Since a short steering duration will generally follow a spiral path with a decreasing radius of curvature, a cornering mark may be found toward the end of this maneuver. A more likely situation is that steering can be *inferred* from a change in direction or a lateral displacement from the "normal" expected travel path.

Determination of the timing of this pre-crash swerve is difficult, particularly in the absence of physical evidence. For a short swerve the lateral acceleration will quickly increase from zero to some maximum value rarely exceeding 0.5

g's on dry pavement. The average lateral acceleration during the sudden swerve would then be limited to approximately 0.25 g's or $a \approx 8$ ft/sec². The time duration of this maneuver would then be given by

$$t \approx \sqrt{2d/a}$$

where

$a \approx 8$ ft/sec²

$t \approx \sqrt{d/4}$

$t \approx \tfrac{1}{2}\sqrt{d}$

where

d = lateral displacement in feet

t = minimum time in seconds.

This is a crude estimate only for the *minimum* time required.

For a long swerve the time required can be determined by using an estimate of the length d of the curved path, which is roughly a segment of a spiral or circle. The time duration of the swerve would then be given by

$$t \approx d/V$$

where

d = swerve path length in feet

V = impact velocity in ft/sec.

Avoidance Braking

Braking is another common avoidance maneuver that may or may not be subject to reconstruction. Physical evidence must be sufficient to determine the braking distance and the appropriate effective drag factor. If this information is known, then the speed at the beginning of this avoidance maneuver can be calculated by consideration of the loss of kinetic energy due to braking.

As discussed in Chapter 3 Vehicle Dynamics, the pre-braking speed is given by the following relationship

$$S_0^2 = S_I^2 + S_B^2$$

or

$$S_0 = \sqrt{S_I^2 + S_B^2}$$

where

S_0 = pre-braking speed
S_I = impact speed
S_B = the equivalent braking speed.

The equivalent braking speed S_B is the speed required to brake to a stop using an effective coefficient of friction f_e over a distance d. That is

$$S_B = 5.47\sqrt{f_e d}\,.$$

The expression $S_o = \sqrt{S_I^2 + S_B^2}$ is basically the combined speed relationship that was discussed earlier.

The time required to execute the pre-crash braking maneuver can be determined from basic kinematics.

$$V_1 = V_0 - at$$

or

$$V_0 = V_1 + at$$

where

V_0 = pre-braking velocity in ft/sec
V_1 = impact velocity in ft/sec ($1.47 S_1$)
t = duration of braking in seconds
a = deceleration rate in ft/sec

and

$$a = f_e g$$

or

$$a = 32.2 f_e.$$

The distance traveled during braking is given by

$$d = V_1 t + \tfrac{1}{2} a t^2. \tag{5.1}$$

Since, d, V_1 and a are all known values, this quadratic equation can be solved for the time t.

For short skids a rough estimate of the braking time can be obtained by using the average speed during braking. That is,

$$V_{ave} = (V_0 + V_1 / 2)$$

and

$$t \approx d / V_{ave}.$$

Acceleration

Accelerating to enter or cross the roadway from a stopped position is a frequent maneuver that may lead to a collision. Assessment of the other driver's reaction to this emergency situation would require a determination of the timing of this acceleration. If we have had a successful reconstruction of the crash to this point, the speed, location and orientation at impact of each vehicle will be known. Referring to Figure 5.2, V2 would have traveled a distance d from a stop and reached a known speed $S2_1$ at impact. From kinematics $a = V / t$ or $t = 1.47 S2_1 / a$.

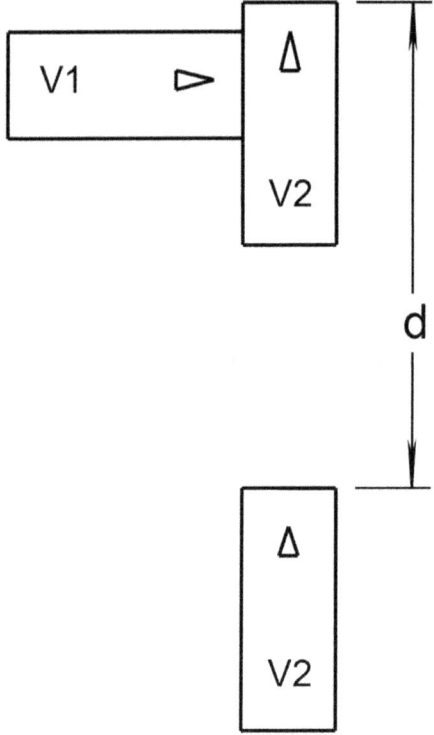

Figure 5.2. Acceleration

An example case is given with the following data:

$S1_1 = 30$ mph = 44 ft/sec
$S2_1 = 10$ mph = 14.7 ft/sec
$d = 25$ ft, $f = 0.8$, and SL (Speed Limit) = 30 mph

No sight distance obstructions

The driver of V#2 (V2-1) said, "I saw V1, but had plenty of time to cross."

The driver of V#1 (V1-1) said, "I was going 30 mph and had no time to avoid."

The time for V2 to reach the POC is equal to

$$t = 2d/V = 2(25)/14.7 = 3.40 \text{ sec}.$$

The acceleration rate of V2 would then be

$$a = V/t = 14.7/3.4 = 4.3 \, ft/sec^2.$$

V2-1's acceleration rate would reflect a normal unhurried maneuver consistent with a failure to see V1 or badly misjudging his speed. A reasonable expectation for V1-1's reaction time in this situation would be approximately 2.0 seconds. (See Chapter 2 Evidence, Reaction Time.) He, therefore, should have started braking at 3.4 - 2.0 = 1.4 seconds prior to impact.

At this time, if traveling at the speed limit of 30 mph, V1 would be a distance d from POC of

$$d = 1.4 \sec(44 ft/sec) = 48 ft.$$

Braking from 30 mph to a stop would require a distance equal to

$$d = S^2/30f = (30)^2/30(.8) = 38 ft.$$

Therefore, he should have been able to stop 10 feet prior to the POC, thus avoiding the impact. It is also possible that he did react promptly and brake but was traveling significantly over the speed limit. As is often the case, the actions of both drivers probably contributed to the causation of the accident.

For a second example we will use the same data except that now $S2_1 = 20$ mph (29.4 ft/sec). As shown in the first example

$$t = 2(25)/29.4 = 1.7 \sec$$

$$a = 29.4/1.7 = 17.3 \, ft/sec^2.$$

This required acceleration is in all probability beyond the vehicle's capability and, therefore, V2 probably "ran" the stop sign.

A third example can demonstrate another use for consideration of a pre-crash acceleration maneuver. In this case we have the same data set except that we are unsure regarding the speed of V2 at impact because of doubts regarding the post-impact departure angle of V1. We have also noted that V1 displays a significant sidesway (see Figure 5.3) to the left from the impact damage that would indicate a likely impact speed for V1 of at least 15 mph. A probable maximum acceleration rate from test data of $a_{max} = 12.5 / \sec^2$ would yield a time to impact equal to

$$t = 2\sqrt{d/a} = \sqrt{2(25)/12.5} = 2.0 \sec$$

and a maximum impact speed of

$$S2_1(\max) = 12.5(2.0) = 25 \, ft/\sec$$

or

$$S2_1(\max) = 17 \, mph.$$

A reasonable estimate of the speed of V1 would, therefore, appear to be approximately 17 mph. It should be recognized that this rough estimate might, or might not, be useful for the analysis of accident causation.

Left Turn

A common collision configuration occurs when a vehicle makes a left turn (right turn in London) in front of an oncoming vehicle. This type of collision is generally difficult to reconstruct since there is rarely any physical evidence on the street (e.g. cornering marks) that would indicate the turning vehicle's direction at impact.

Damage patterns are generally sufficient to determine if the vehicle was turning or merely crossing the intersection. These damage patterns do allow for a broad estimate of the relative orientation of the vehicles and, therefore, a similarly broad estimate of the turning vehicle's approach direction. Laying out a "logical" turning path leading to the impact configuration can enhance this estimate. Sometimes this exercise can yield a reasonable estimate of the range of speed for the striking vehicle. It is, however, less likely that the corresponding speed estimate for the turning vehicle will be reliable. An assumption of "normal" lateral friction demand may lead to a speed estimate for the turning vehicle that is as valid as our convoluted reconstruction.

The timing of pre-crash events can be estimated using the calculated speeds from the reconstruction effort. The timing can then be used to evaluate the behavior of each driver. That is, should the vehicle have turned and could the striking vehicle have been expected to avoid the collision. These conclusions should receive careful scrutiny since the basis may not be fully justified.

Some crash configurations allow for simplified solutions but still, however, rely upon the same principles utilized in the analysis of the general intersection type collision.

Collision Configurations

Broadside Collisions

In this case the striking vehicle V1 impacts the stopped vehicle V2 in the side between the axles, and pushes it sideways to final rest. V2 will be in a full side skid and assuming no braking V1 will be rolling. Since there will be no separation and the impact forces will not go to zero, the impact will essentially continue to rest. A simple way to analyze the "post-impact" phase of this collision is to consider the two vehicles as an eight-wheeled composite vehicle with only four wheels braking. The effective friction factor f_e for this partial braking is proportional to the load distribution on V2. That is

$$f_e = \frac{W_2}{W_1 + W_2} f$$

where f is the coefficient of friction (or drag factor) for a skidding tire. The composite post-impact speed SC_2 would be

$$SC_2 = 5.47\sqrt{f_e d}$$

where

$$SC_2' = S1_2 = S2_2.$$

If the front tires of V1 are jammed during the impact, this eight-wheeled composite vehicle will have six wheels braking. The effective coefficient of friction would then be given by

$$f_e = \frac{W_2 + W_{1F}}{W_1 + W_2} f$$

where

W_{1F} = weight of V1 on the front wheels.

Determination of the impact speed may be achieved by two separate methods. From conservation of momentum

$$S1_1 W_1 + S2_1 W_2 = S1_2 W_1 + S2_2 W_2.$$

Since $S2_1 = 0$ and $SC_2 = S1_2 = S2_2$, this would reduce to

$$S1_1 W_1 = SC_2 (W_1 + W_2)$$

or

$$S1_1 = \frac{W_1 + W_2}{W_1} SC_2. \tag{5.2}$$

A second method would be to combine the post-impact kinetic energy with the energy lost during impact to find the total kinetic energy at impact. Since the damage patterns produced by this crash configuration are generally similar to those produced by staged collisions, this method should be reasonably reliable.

Using the relationships developed from test results, the fixed barrier equivalent FBE speed S_B would be given in the form described in Chapter 4 Impact Dynamics

$$S_B = b_0 + b_1 C$$

where C is the damage crush.

The kinetic energy loss for each vehicle during impact due to damage would then be equal to

$$KE_D = \frac{W}{2g}(1.47 S_B)^2.$$

The total kinetic energy loss KE_D during impact would be given by

$$KE_D = \frac{W_1}{2g}(1.47 S1_B)^2 + \frac{W_2}{2g}(1.47 S2_B)^2$$

where

$S1_B$ = fixed barrier equivalent speed (FBE) of V1
$S2_B$ = fixed barrier equivalent speed (FBE) of V2.

The total kinetic energy KE_P loss post-impact is equal to

$$KE(\text{post}) = \frac{W_1 + W_2}{2g}(1.47 SC_2)^2$$

where SC_2 is the post impact speed of the composite vehicle and is equal to

$$SC_2 = S1_2 = S2_2.$$

Since the impact kinetic energy is given by

$$KE(\text{Impact}) = \frac{W_1}{2g}(1.47 S1_1)^2.$$

Equating the kinetic energy at impact to the sum of the energy loss during impact plus the post-impact energy loss yields:

$$\frac{W_1}{2g}(1.47 S1_1)^2 = \frac{W_1}{2g}(1.47 S1_B)^2 + \frac{W_2}{2g}(1.47 S2_B)^2 + \frac{W_1 + W_2}{2g}(1.47 SC_2)^2.$$

Simplifying yields

$$W_1 S1_1^2 = W_1 S1_B^2 + W_2 S2_B^2 + (W_1 + W_2) SC_2^2. \tag{5.3}$$

Since W_1, W_2, $S1_B$, $S2_B$, and SC_2 are known values, solving equation (5.3) yields the impact speed $S1_1$ of V1 without directly calculating any energy values.

An alternate approach is to calculate the actual values in ft-lb of energy loss during impact and post-impact energy. Summing these yields

$$KE_I = KE(\text{Impact}) = KE(\text{Damage}) + KE(\text{Post}).$$

Since

$$KE_I = \frac{W_1}{2g}(1.47 S1_1)^2$$

$$S1_1^2 = \frac{2(32.2)KE_I}{(1.47)^2 W_1}$$

or

$$S1_1^2 = \frac{30}{W_1} KE_I$$

and

$$S1_1 = \sqrt{5.47 KE_I / W}. \qquad (5.4)$$

The work energy method, using vehicle damage, is generally not as reliable as that given by conservation of energy in Equation (5.2), but it does provide a valuable cross check for gross errors.

Head-On Collisions

A head-on collision can be reconstructed in a manner similar to other two-vehicle accidents except the energy losses due to damage often must be taken into consideration. An off-center collision, as shown in Figure 5.3, will produce post-impact dynamics that will include yawing, partial braking if any front tires are jammed, rolling, and in some cases, a rollover.

ANALYSIS PROCEDURES | 127

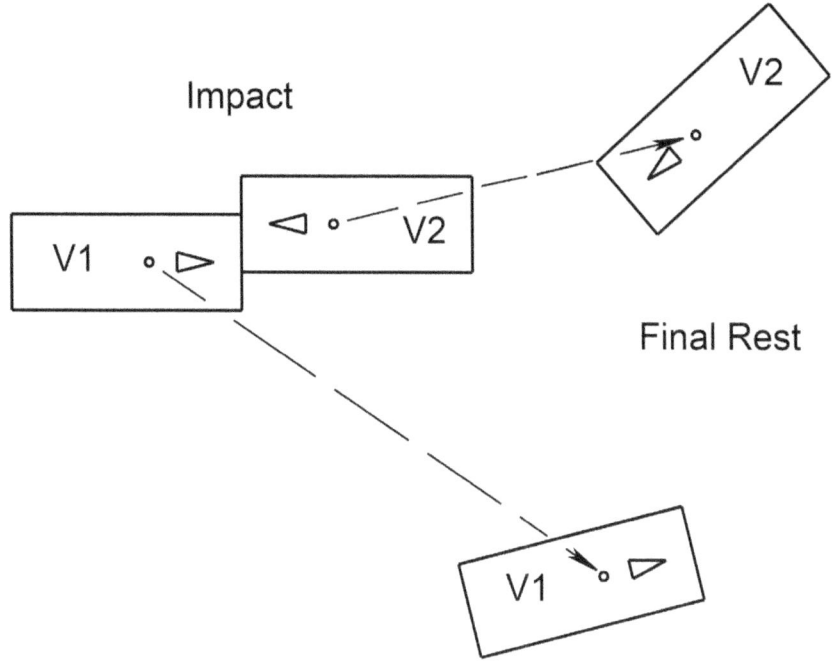

Figure 5.3. Head-On Collision

Calculation of the post-impact speeds can be done as described in Chapter 3 Vehicle Dynamics. Determination of the departure direction must, as always, come from physical evidence. Since the changes in direction are usually quite small in this crash configuration, estimates of the lateral post-impact speeds are often of questionable reliability. Therefore, the use of $S1_2(y)$ or $S2_2(y)$ in a subsequent conservation of momentum analysis is rarely appropriate. We now have an impact analysis with two unknown approach speeds $S1_1$ and $S2_2$ and only one independent momentum equation.

$$S1_1(x)W_1 + S2_1(x)W_2 = S1_2(x)W_1 + S2_2(x)W_2. \qquad (5.5)$$

In this crash configuration the approach or impact speeds are

$$S1_1 = S1_1(x) \text{ and } S2_1 = S2_1(x).$$

Since this is generally an inelastic impact, the post-impact speeds are approximately equal. That is

$$S1_2(x) \approx S2_2(x) \approx S_2(x).$$

Equation (5.5) now reduces to

$$S1_1 W_1 + S2_1 W_2 = S_2(x)(W_1 + W_2). \qquad (5.6)$$

Using damage test data and the crush to each vehicle the FBE speed S_B for each vehicle is as previously shown equal to

$$S_B = b_0 + b_1 c \text{ or } S_B = b_1 c.$$

Note that these speeds must be adjusted for a partial crush profile as shown in Chapter 4 Impact Dynamics. The kinetic energy loss due to damage can be expressed as

$$KE(\text{Damage}) = \frac{W_1}{2g}(1.47 S1_B)^2 + \frac{W_2}{2g}(1.47 S2_B)^2.$$

The post-impact kinetic energy is given by

$$KE(\text{Post}) = \frac{W_1 + W_2}{2g}(1.47 S_2(x))^2.$$

The impact kinetic energy is equal to

$$KE(\text{Impact}) = \frac{W_1}{2g}(1.47 S1_1)^2 + \frac{W_2}{2g}(1.47 S2_1)^2.$$

Since KE(Impact) = KE(Damage) + KE(Post), we have the following equation

$$\frac{W_1}{2g}(1.47 S1_1)^2 + \frac{W_2}{2g}(1.47 S2_1)^2$$
$$= \frac{W_1}{2g}(1.47 S1_B)^2 + \frac{W_2}{2g}(1.47 S2_B)^2 + \frac{W_1 + W_2}{2g}(1.47 S_2(x))^2.$$

Simplifying yields

$$W_1 S1_1^2 + W_2 S2_1^2 = W_1 S1_B^2 + W_1 S2_B^2 + (W_1 + W_2) S_2^2. \qquad (5.7)$$

Substituting the known values W_1, W_2, $S_2(x)$, $S1_B$, and $S2_B$ equations (5.6) and (5.7), and solving them simultaneously yields the solution of the approach speeds $S1_1$ and $S2_1$ of V1 and V2. Although this analysis appears cumbersome, after actually substituting the known values into equations (5.6) and (5.7), this becomes a rather simple simultaneous solution. If there is full frontal impact, there will be no lateral movement, thus simplifying the post-impact analysis. The damage analysis procedure will also be simplified since the crush profile will be constant over the full vehicle width. The procedure, however, will be the same as that used for the off-center head-on collision.

In some head-on collisions there is no post-impact movement. That is

$$S1_2 = S2_2 = 0.$$

This "perfect" head-on collision can be treated as two separate fixed barrier impacts and solved independently. Calculation of the impact speeds, $S1_1$ and $S2_1$, is carried out in the same fashion as shown in Chapter 4 Impact Dynamics. Each impact speed would be given by one of the two following relationships.

$$S_1 = b_1 c \text{ or } S_1 = b_0 + b_1 c.$$

In this case conservation of momentum provides a simpler and more reliable solution since the total momentum after impact is equal to zero

$$S1_1 W_1 - S2_1 W_2 = 0$$

or

$$S1_1 = \frac{W_2}{W_1} S2_1$$

or

$$\Delta V_1 = \frac{W_2}{W_1} \Delta V_2.$$

Conservation of momentum is obviously the preferred method in this case, but the damage-only option provides a reasonable cross check.

Rear-End Collisions

This crash configuration may be solved in the same manner as the head-on collision. The departure speeds for both vehicles will be approximately the same but the striking vehicle V1 must be departing at a speed less than or equal to that of the struck vehicle. That is

$$S1_2 \leq S2_2.$$

As in the head-on collision, kinetic energy losses during impact and in the post-impact motion must be utilized in addition to conservation of momentum. In general this solution is less reliable since crush data for rear impacts is limited.

If the struck vehicle is stopped at impact $(S2_1 = 0)$, then conservation of momentum is often sufficient for a reasonable impact analysis. Assuming that $S1_2 \approx S2_2 \approx SC_2$, the momentum relationship yields one equation in one unknown

$$S1_1 W_1 = SC_2(W_1 + W_2)$$

or

$$S1_1 = \frac{W_1 + W_2}{W_1} SC_2.$$

Since this last case is a fairly common occurrence, the necessity for utilizing rear impact damage can be avoided in the analysis of many rear-end collisions. Note that for a low speed collision this impact will not be totally inelastic, separation will occur, and $S1_2$ and $S2_2$ are not equal. A solution will require a full post-impact analysis, a calculation of the FBE speeds S_B, and a solution similar to that used for the head-on collision.

Chain Reaction Rear-End Impacts

Theoretically, this rather common collision event could be solved if all points of impact and all distances are known and we have good crush measurements and reliable crash test data. In practice a reliable solution is highly improbable since pre-crash speeds of the struck vehicles are usually unknown, crash test data is often lacking, and the sequencing of the impacts cannot always be determined. That is, the question of who hit whom first cannot be answered.

If the rear-most vehicle V1 initiates an impact sequence with a line of stopped vehicles, each impact would possibly be amenable to a solution using conservation of momentum only. For a three-vehicle collision, the relationships would take the following form

$$S1_1 = \frac{W_1 + W_2}{W_1} S2_2 \qquad (5.8)$$

assuming that $S1_2 = S2_2$

$$S2_3 = \frac{W_2 + W_3}{W_2} S3_4 \qquad (5.9)$$

assuming that $S2_4 = S3_4$. It is also assumed that the impacts are closely spaced so that $S2_2 = S2_3$. Substituting the final post-impact speed $S3_4$, found from a post-impact analysis, Equation (5.9) yields $S1_1$ and $S2_2$ which then, from Equation (5.8) yields the speed of the striking vehicle $S1_1$.

In many instances we are more interested in determining the qualitative reconstruction rather than precise speeds, that is, which vehicle started the chain reaction. One useful observation is that, if the rear-most vehicle initiates the impact, you would expect the damage to diminish as you move down the stream.

Underride/Override Collisions

A rather typical rear-end collision is preceeded by the striking vehicle V1 braking with the resulting dipping of the vehicle front. V1's front bumper then underrides the rear bumper of V2, thus causing extensive and expensive damage to soft parts such as the grill, lights and fenders of V1.

Although this may appear as poetic justice for tailgating and/or inattention, it creates a problem for any reconstruction effort. The use of crush damage to estimate kinetic energy loss is severely compromised since the determination of an effective stiffness coefficient is no better than an educated guess. If, however, conservation of momentum is sufficient for the analysis of the impact, then this type of impact does not pose a problem. An example of a pickup truck underride of the side of a tractor-trailer is shown in Figure 5.4. The redirection gouges from the undercarriage of the pickup, as shown in Figure 5.5, indicate the movement of the tractor-trailer during impact. A severe rear-end override of a truck onto a passenger car is shown in Figure 5.6. A determination of the vehicle change in speed due to either of these collisions would essentially be based upon a SWAG[4].

Figure 5.4. Underride Collision

---

[4] Scientific wild ✳✳✳ guess

Figure 5.5. Redirection Gouges

Figure 5.6. Override Collision

Sideswipes

Sideswipe damage such as gouges, scrapes, paint peels and displacement of vehicle components reveals the relative motion of vehicles and possibly the points of contact. It does not provide much useful help in speed determination. Note that this impact configuration is rarely a candidate for a computer solution.

Multiple Impacts

Intersection collisions often result in the post-impact trajectory terminating by impact with a pole, tree, building, parked vehicle or other roadside object. Keeping in mind that in the post-impact analysis we are looking at losses in kinetic energy, this does not pose a serious problem.

This final impact can be treated as a single vehicle impact by computing an estimated energy loss through a damage analysis. Adding this energy loss to that loss during the post-impact rollout, spinout or rollover yields the total post-impact kinetic energy. This relationship is essentially the same as the combined speed relationship for braking prior to an impact. This can be expressed as

$$S_2^2 = \overline{S^2} + S_3^2$$

where

S_2 = post-impact speed
S_3 = final impact speed
$\overline{S}$ = post-impact speed ignoring the final impact.

When the final impact speed is relatively small, poor estimates of this speed will have little effect on the post-impact speed computation.

Truck-Car Collisions

Although most of the previous discussions in this chapter have been limited to collisions between passenger cars, the principles involved are applicable to accidents involving all types of vehicles and objects. The analysis of the impact phase of a collision between a large truck and an automobile can, however, be difficult.

Using conservation of momentum to analyze an intersection impact is hampered by the weight disparity between the two vehicles. Changes in direction of the truck are small, therefore, small errors in the estimation of its departure angle will lead to large errors in the estimated speed of the car.

If the impact configuration requires the use of crush damage in the analysis, we are limited since there are virtually no stiffness criteria for any part of the truck. If the truck damage is essentially zero, it may be considered a moving rigid barrier. An example of this would be a head-on collision where all of the kinetic energy lost during impact would be due to crush damage to the car. The remainder of the analysis would be the same as the previously described head-on collision between two cars.

Truck-Truck Collisions

Unless conservation of momentum can yield a reliable impact analysis, truck-to-truck collisions are generally limited to a qualitative evaluation. Physical evidence on the street and vehicle damage is usually sufficient to establish the crash configuration.

Multiple Vehicle Collisions

Collisions involving dozens of vehicles that often occur in smoke and/or fog leave an aftermath that creates the ultimate reconstruction challenge. At best we might, if present at the actual scene, be able to find evidence sufficient to qualitatively describe a few of the individual impacts. I was asked why I could not completely reconstruct the accident shown in Figure 5.7 and my reply was that I was an engineer, not a wizard.

136 | HIGHWAY ACCIDENTS

Figure 5.7. Multiple Vehicle Collisions

Car-Motorcycle Collisions

Collisions between cars and motorcycles cannot generally be reconstructed quantitatively due to the weight differential and the lack of any significant crash data for motorcycles. Any change in direction of the car is difficult to quantify and, therefore, is of little value in estimating the speed of the motorcycle.

Common impact configurations are the "T-bone" when a car pulls out in front of a motorcycle and an angled collision in which a car turns left in front of a motorcycle. The damage to the turning vehicle, see Figure 5.9, does not allow for any meaningful speed estimate except that it does not demonstrate a high-speed impact. The damage to the front of the right front door indicates that the motorcyclist may have started a right swerve avoidance maneuver just prior to impact. The crush to the rear of the right rear door clearly shows the car was moving at impact, thus causing a clockwise rotation of the motorcycle, leading to this rather hard secondary sideslap. These type of crashes are usually caused by the car's driver, but the motorcyclist might also have been speeding or inattentive.

Figure 5.8. Car – Motorcycle Collision

In such crashes speed estimates using the car damage are also suspect since the crush pattern will be narrow and vertical which generally does not match any test data. Any speeds calculated should be considered crude estimates.

Pedestrian-Vehicle Collisions

When a pedestrian is struck by a van, bus or truck, the pedestrian and the vehicle have approximately the same longitudinal post-impact departure speed. If the vehicle is not braking, the pedestrian will fall to the ground, decelerate and be run over. If the vehicle is braking, the pedestrian will roll/tumble/slide to final rest and the vehicle will generally stop prior to reaching the pedestrian. In this case determination of the speed of the vehicle may be found by calculating the post-impact speed of the pedestrian. A rough estimate of the pedestrian's post-impact speed SP_2 can be found by using the relationship

$$SP_2 = 5.47\sqrt{fed}$$

where

fe = effective drag factor for this roll/tumble/slide
d = distance from impact to final rest.

Determination of *fe*, which is approximately within the range of 0.3 to 0.6, is difficult since, for obvious reasons, testing has been limited to the use of dummies. This analysis, of course, requires that the impact point can be established through physical evidence (e.g, a. shoe scuff).

When a car strikes an adult pedestrian, the body may fold over onto the hood and then may strike the windshield. If the speed of the car is sufficient, the body may continue up and onto the roof and then fall to the roadway as the car passes under it. If the vehicle is braking at impact, the body may be propelled forward and upward without a secondary impact and follow a parabolic trajectory to landing on the ground. The body will then roll/tumble/slide to final rest.

An attempt to apply the vault concept (projectile motion) to this pedestrian motion is unlikely to be successful. Locating the point of impact by physical evidence (e.g., shoe scuffs) is rare. A general location such as a crosswalk may provide the basis for a hypothetical solution. Determining the touch down point may also be quite difficult since there is often no blood, tissue or other evidence at this point. Without both of these locations no reasonable estimate of the vault distance can be made.

If d is known and we assume a reasonable estimate of h (~2 to 3 feet), we still have in almost all cases a totally unknown take off angle θ_1. If the touch down and final rest are known, then this post-crash roll, tumble and slide can probably be used to give the best available, but not highly accurate, estimate of the body's horizontal speed V_{1X} post-impact. This speed, however, will be somewhat less than the speed of the car at impact.

In many cases the major issue is not the speed of the vehicle, but whether the pedestrian was stopped, walking normally ($\pm$ 5 ft/sec) or running prior to impact. As shown in Figure 5.9, if the departure direction (dashed line) is line A or B with $\theta \approx 0$, then the pedestrian must have been stopped or walking slowly.

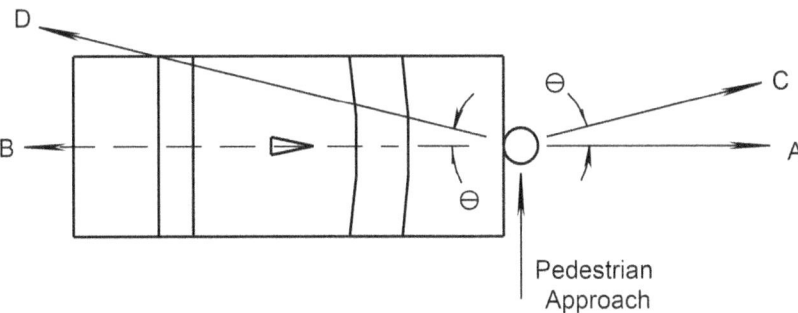

Figure 5.9. Pedestrian – Car Impact

If the points of impact and final rest for the pedestrian are known, the departure angle θ as given by path C can be found. This angle θ can also be found from the path D across the hood of a car without knowing either the impact point or the final rest position of the pedestrian. In either case the relative speed of the pedestrian and the car is given by the relationship

$$SP_1 \geq S1_1 \tan \theta \tag{5.10}$$

where

SP_1 = pre-crash speed of the pedestrian
$S1_1$ = pre-crash speed of the vehicle.

As expressed in Equation (5.10), SP_1 may be somewhat greater due to the friction force created by the pedestrian's forward motion across the front of the vehicle.

A car or motorcycle occupant ejection during a collision is difficult to analyze for the same reasons that apply to pedestrian impacts. In addition the occupant may experience a dramatic and unknown reduction in speed during the ejection process.

Bicycle-Vehicle Collisions

A bicycle impact can be treated in the same fashion as a pedestrian impact. However, the post-impact trajectory of the bicycle is erratic and cannot be used with any reliability.

Single Vehicle Accidents

Single vehicle accidents represent approximately 20% of the total accidents, but account for about 60% of fatal crashes and 40% of the total number of fatalities. They are, therefore, an important concern in the analysis of causal factors and the development of counter measures.

Some single vehicle events, such as a spinout with or without a subsequent rollover, or a truck jackknife may take place on the pavement. However, most single vehicle accidents involve roadway departures which may be due to a vehicle problem, an avoidance maneuver, excessive speed, or driver inattention.

The roadway departure is generally followed by a braking action, a spinout, a yaw with a subsequent rollover, an impact with a roadside object or a combination of these. There also may be some further movement after striking a roadside object.

Determination of the initial speed can be accomplished by following the previously discussed procedures. The energy loss in each phase of the accident event is calculated as shown in Chapter 3 Vehicle Dynamics and Chapter 4

Impact Dynamics. Summing these energy losses then yields the total kinetic energy and the vehicle speed.

Motorcycles not only leave the roadway and roll over or impact roadside objects, but also roll over due to loss of control caused by one of the following:

- Unpaved roadways
- Open mesh steel grids on bridges
- Surface contaminants, such as oil
- Ice or snow
- Pot holes
- Raised manhole covers
- Uneven pavement
- Pavement shoulder drop-offs
- Avoidance maneuvers
- Excessive speed, particularly in combination with turning or braking
- Reckless maneuvers (e.g. wheelies)

Bicycles can encounter similar problems and can also face additional problems on bridges such as open mesh steel grids and gaps between adjacent bridge spans that can bring the bicycle (not the rider) to a sudden stop. See Figure 5.10.

Figure 5.10. Bicycle – Bridge Encounter

Critical Speed

A vehicle following a curved path can suffer a loss of control due to a rollover or a spinout. These two failure mechanisms can theoretically be used to calculate the speed of the vehicle. In both cases the centripetal acceleration is given by

$$a = V^2 / R$$

(See derivation in the Appendix Physics Review, Circular Motion.)

where

a = centripetal acceleration in feet/second^2
V = vehicle tangential velocity in feet/second
R = path radius in feet.

ANALYSIS PROCEDURES | 143

The centrifugal inertial F_I force tending to push the vehicle off the path is, therefore, equal to

$$F_I = \frac{W}{g} \frac{V^2}{R}$$

where

F_1 force in lbs
W = vehicle weight
g = 32.2 ft/sec².

Rollover

An example of a vehicle with a high center of gravity that would be prone to a rollover rather than a spinout would be the standard gasoline tanker truck. There have been an alarming number of spectacular crashes involving tankers fully loaded, traveling too fast on curved freeway entrance and exit ramps.

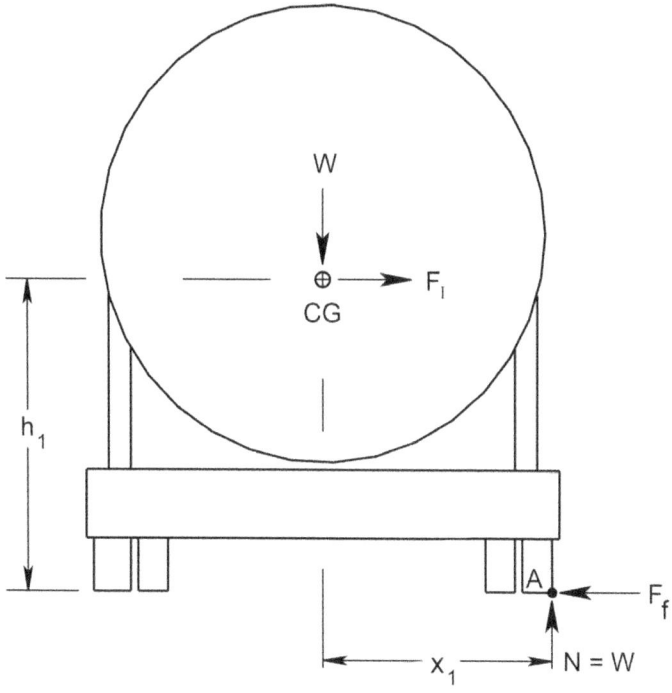

Figure 5.11. Tanker Rollover

As shown in Figure 5.11, h_1 is the height of the center of gravity CG and x_1 is the horizontal distance from the CG to the outer edge of the outside tire. For equilibrium the centrifugal force F_t is equal to the friction force F_f acting through point A and the normal force N is equal to the vehicle weight W. At impending tipping the summation of the moments about point A will be equal to zero. That is

$$\sum M_A = F_t[h_1] - W[x_1] = 0$$

or

$$\frac{W}{g}\frac{V^2}{R}h_1 = Wx_1$$

or

$$V^2 = gRx_1/h_1$$

and

$$V = 5.67\sqrt{Rx_1/h_1}$$

or

$$S = 3.87\sqrt{Rx_1/h_1}. \qquad (5.11)$$

This would then be the theoretical speed of the truck at rollover. If the tanker truck is a tractor-trailer combination, Equation (5.11) will give the speed of the lift-off of the inside trailer tires but the resistance of the tractor will slightly delay the complete rollover. As the trailer continues to rotate or some small change in steering occurs, the tractor will be forced into a complete rollover. There are, however, several problems affecting this analysis, including:

1 The height of the center of gravity h_1 may be unknown.
2 The required adjustments to h_1 and x_1 as described in Chapter 3 Vehicle Dynamics "Initiation of the Rollover," are difficult to estimate.
3 The path of the truck may not coincide with the highway curve – cornering marks are not always present.

4 An undetected sudden steering input or braking action may precipitate the rollover.

5 A resulting fire may destroy much of the vital evidence.

In general the use of equation (5.11) to estimate the speed at rollover should be viewed with considerable skepticism.

Spinout

Vehicles with a low center of gravity will spinout prior to reaching a rollover speed. When the centrifugal inertial force F_I exerted on the vehicle exceeds the available centripetal friction force F_f, a spinout will occur with the vehicle leaving the intended circular path. As in an impending rollover, the centrifugal force is equal to

$$F_I = \frac{W}{g}\frac{V^2}{R}. \tag{5.12}$$

The restraining friction force is given by

$$F_f = fW. \tag{5.13}$$

Equating (5.12) and (5.13) yields

$$V^2 = gfR$$

since

$$V = 1.47S$$

$$S^2 = 15fR$$

and

$$S = 3.87\sqrt{fR}$$

where

> S = theoretical speed of spinout
> f = maximum lateral coefficient of friction
> R = radius of the travel path.

During a "high speed" cornering maneuver the weight is transferred to the outer edge of the outside tires. This can produce a "squealing" noise and/or leave a narrow black cornering tire mark. Since this physical evidence can be seen (or heard) at speeds well below impending spinout, the use of the critical speed formula can significantly overestimate the vehicle speed.

In many cases there is no physical evidence prior to the spinout and, therefore, the vehicle path radius is unknown. It is not necessarily the same as the highway curve and without supporting evidence the loss of control cannot be directly related to the speed and curve radius.

In negotiating a curve, a driver executes normal course corrections simply to maintain the desired travel path. These steering inputs significantly change the instantaneous radius and the "critical" speed. Most drivers cannot maintain a curved path that requires a lateral coefficient of friction greater than 0.5.

Spinouts can also be caused by a sudden steering input that would drastically reduce the path radius or by a braking action that would suddenly reduce the available lateral friction capacity. These driver actions, generally undetectable, would make the critical speed formula totally inappropriate.

In some cases there will be a clear cornering mark becomes darker and decreases in radius as the vehicle approaches spinout. It is possible that this is evidence of critical speed loss of control, but a problem arises in measuring the actual radius at the termination of the cornering mark. Using the method of measuring a chord and middle ordinate will yield an average radius of curvature rather than the smaller instantaneous radius oat the termination of the cornering mark. This value for the radius, that is too large, will yield a speed estimate that is too high. Certain advanced surveying equipment and computer techniques can provide an improved estimate of the value of the terminal radius of curvature.

Another serious, but all too common, misuse of this formula is applying it to the path of a yaw mark. The tire yaw mark is *not* traveling on the same path as the vehicle CG, is *not* a cornering mark and has nothing to do with the cause of the spinout. Therefore, using the critical speed concept is totally invalid.

In summary the critical speed formula is probably the most misused method for estimating the speed of a vehicle at spinout. I have rarely had a situation where I felt it could be used with any confidence in the reliability of the results. I feel we would be better served using the methodology described in Chapter 3 Vehicle Dynamics to estimate the vehicle speed at the initiation of the spinout. This spinout analysis should at the least be used as a check on the results obtained from the critical speed formula.

Vault

When a vehicle is airborne during its travel path, such as shown in Figure 5.12, this is generally called a vault. The vehicle begins the vault with a velocity V_1, travels horizontally a distance d, and drops a distance h to it's landing point. A generalized diagram, illustrating this projectile motion, is shown in Figure 5.13. The equation for this motion is given by

$$y = \tan\theta_1 x - \frac{g}{2V_1^2 \cos^2\theta_1} x^2. \qquad (5.14)$$

The derivation of equation (5.14) is given in the Appendix Physics Review, Projectile Motion.

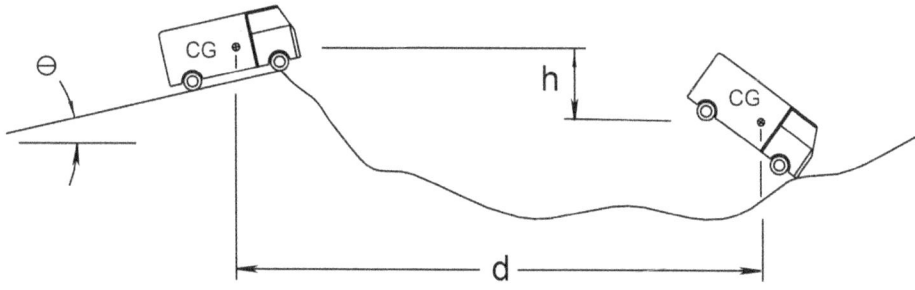

Figure 5.12. Vault

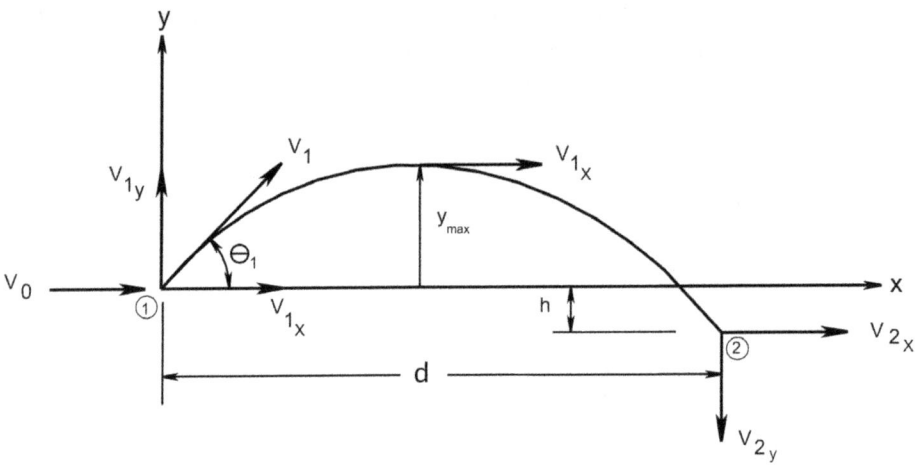

Figure 5.13. Vault

For reconstruction purposes, if θ_1, h and d are known values, as shown in Figure 5.13, then we can solve for the velocity V_1 at point (1). At $x = d$ and $y = -h$, Equation (5.14) becomes

$$-h = \tan\theta_1 d - \frac{gd^2}{2V_1^2 \cos^2\theta_1}$$

or

$$\tan\theta_1 d + h = \frac{gd^2}{2V_1^2 \cos^2\theta_1}.$$

Solving for V_1^2 yields

$$V_1^2 = \frac{gd^2}{2\cos^2\theta_1(\tan\theta_1 d + h)}$$

and

$$V_1 = d\sqrt{\frac{16.1}{\cos^2\theta(\tan\theta_1 d + h)}}. \tag{5.15}$$

If the landing point (2) is above point (1), then h will be a negative value.

The use of Equation (5.15) to find a realistic estimate for a vehicle's take off velocity V_1 is dependent upon reasonable estimates of h, d and θ_1. Not only must the take off and touch down points be known, but also the vehicle orientation at each point must be known since h and d are the measurements of the change in location of the center of gravity of the rotating vehicle.

In most cases (for example, a flip preceding a rollover) the take off angle θ_1 is not known. Estimates of this angle are generally not much better than guesses.

There is still a problem even if you are able to determine what you feel is a usable estimate of θ_1, and therefore V_1. Since there will be a certain loss of kinetic energy in the vaulting mechanism, the approach velocity V_0 will not be equal to V_1 or V_{1X}. All that we can be assured of is that V_0 is greater than V_{1X} by some unknown amount.

Since this method of velocity determination using Equation (5.15) may be based upon dubious assumptions, its use is rarely justified. In fact, after "critical speed" this is probably the second most misused concept. However, since little or no kinetic energy is lost during the vault itself, this method can often be avoided.

A particular case that may yield a reasonable solution is a vault caused by a vehicle leaving the roadway at a known angle such as shown in Figure 5.12. This might occur in hilly terrain or exiting on a super-elevated curve. The use of Equation (5.15) may yield a reasonable solution since the exit angle θ_1 can be determined, and the approach velocity V_0 is equal to V_1. The slope θ_1 usually must be measured at the accident site since it often cannot be obtained from the highway plans. A reasonable estimation of the vehicle orientation at its landing point is required since d and h are measurements to the vehicle center of gravity.

Although not necessary, another simplification of Equation (5.15) may be made for this vault configuration. Take off angles are shallow since roadway grades and cross slopes rarely exceed 10 percent ($\theta = 5.7$ degrees). For small angles $\cos^2 \theta \approx 1.0$ and $\tan \theta =$ grade g of the exit path. The simplified form of the equation is now

$$V_0 = d \sqrt{\frac{16.1}{(gd + h)}} \ . \qquad (5.16)$$

Another common simplification of equation (5.15) is used to determine the minimum velocity V_1. This is accomplished by assuming that $\theta = 45°$ and, therefore, $\tan\theta = 1.0$ and $\cos^2\theta = 0.50$. Substituting into Equation (5.15) yields

$$V_1 = d\sqrt{\frac{32.2}{d+h}}$$

or

$$V_1 = 5.67d\sqrt{\frac{1}{d+h}} \ . \qquad (5.17)$$

This is not, of course, equal to V_0 which is generally greater than V_1. Although Equation (5.17) does not yield an actual speed estimate, the result may be of some interest and may be considered as providing a reconstruction boundary. Expressing Equation (5.17) in terms of speed yields

$$S_1 = 3.87d\sqrt{\frac{1}{d+h}} \ . \qquad (5.18)$$

A final case using the concept of projectile motion is a vehicle driving horizontally off a cliff or a building. In this case $V_0 = V_{1X} = V_1$, $\theta = 0$, $\cos^2\theta = 1.0$, and $\tan\theta = 0$, reducing Equation (5.15) to

$$V_0 = d\sqrt{\frac{16.1}{h}} \ .$$

For example, if h = 64 ft d and d = 44 ft

$$V_0 = 44\sqrt{\frac{16.1}{64}}$$

or

$$V_0 = 22 \text{ ft/sec} = 15 \text{ mph} \ .$$

An alternative solution is to use simple kinematic relationships. The vertical drop h is equal to

$$h = \frac{1}{2}gt^2$$

or

$$t = \sqrt{\frac{2h}{g}}.$$

For our example the time to drop is equal to

$$t = \sqrt{\frac{2(64)}{32}} = 2.0 \text{ sec}$$

since $V_X = V_0 = d/t$

$$V_0 = 44/2 = 22 \text{ ft/sec} = 15 \text{ mph}.$$

Again we must remember that h and d are measurements of the displacement of the CG of the vehicle.

Summary

At the conclusion of the reconstruction effort an assessment of the results should be carried out. Whenever possible, a cross check of each calculation should be conducted using an alternative procedure. Reconstruction is not an exact science. Many assumptions and educated estimates are used throughout the process. A sensitivity analysis can be conducted by using a range of different values for such variables as stiffness coefficients, effective drag factors and, especially, approach and departure angles. This sensitivity analysis can be greatly simplified by the use of rather simple computer programs. Keep in mind at this stage the computer is doing exactly the same as you would with a hand calculator, but a whole lot faster. The use of computers in reconstruction will be discussed in more detail later.

In the presentation of quantitative conclusions it is important to realize that the number of significant figures you use provides an implicit expression of the degree of accuracy of your solution. For example, if my calculated speed is given as 32.73 mph I am in effect saying that this speed is accurate to ± .01 mph. Reporting more than two significant figures (e.g., 33 mph) is rarely justified although it is not unreasonable to use more significant figures for intermediate computations. Giving a speed estimate as a range (e.g., 30 to

35 mph) or a given value with an estimate range of error (e.g., ±5 mph) is probably the most appropriate procedure.

In many cases a complete reconstruction is not possible, but some of the quantitative reconstruction conclusions may be useful. In some cases you may be able to obtain a reliable speed for only one of the vehicles or you may only be able to get a rather large range of speed for a vehicle. This limited information may, however, be useful for your purposes. For example, if your best effort in determining a vehicle's speed is $30 < S1_1 < 50$, this may be quite adequate if the speed limit is 15 mph.

It is also necessary to remember that the qualitative reconstruction is more important than any quantitative conclusions. That is, knowing what happened is more important than knowing how fast it happened. Finally, sometimes a reasonable and easily defended conclusion when asked what happened is to say "I don't know."

Examples 6

This chapter provides specific example reconstructions of accidents with a variety of configurations and different analysis procedures. Three intersection type collisions included are:

1. A car vs. car intersection collision with a complete quantitative reconstruction analysis.
2. An intersection collision between a truck and an SUV with a complete reconstruction and a complete causal analysis.
3. A qualitative reconstruction of a car vs. car intersection collision with a partial causal analysis.

Also addressed are the analyses of several off-roadway accidents that include: a rollover; impacting a barrier, an energy-absorbing device, and a breakaway pole; and entering a pavement/shoulder drop-off.

Car vs. Car Intersection Collision – Quantitative Analysis

This example intersection collision involves two passenger cars colliding as shown in Figure 6.1. For the sake of simplicity, the roadway edges and pavement markings are not shown. Side scuffs from V2, an impact redirection mark from the left front tire of V1 and the vehicle damage patterns establish the location and orientation of each vehicle at impact. Each vehicle appears to be in its proper travel path. Therefore, there was probably no significant pre-crash steering.

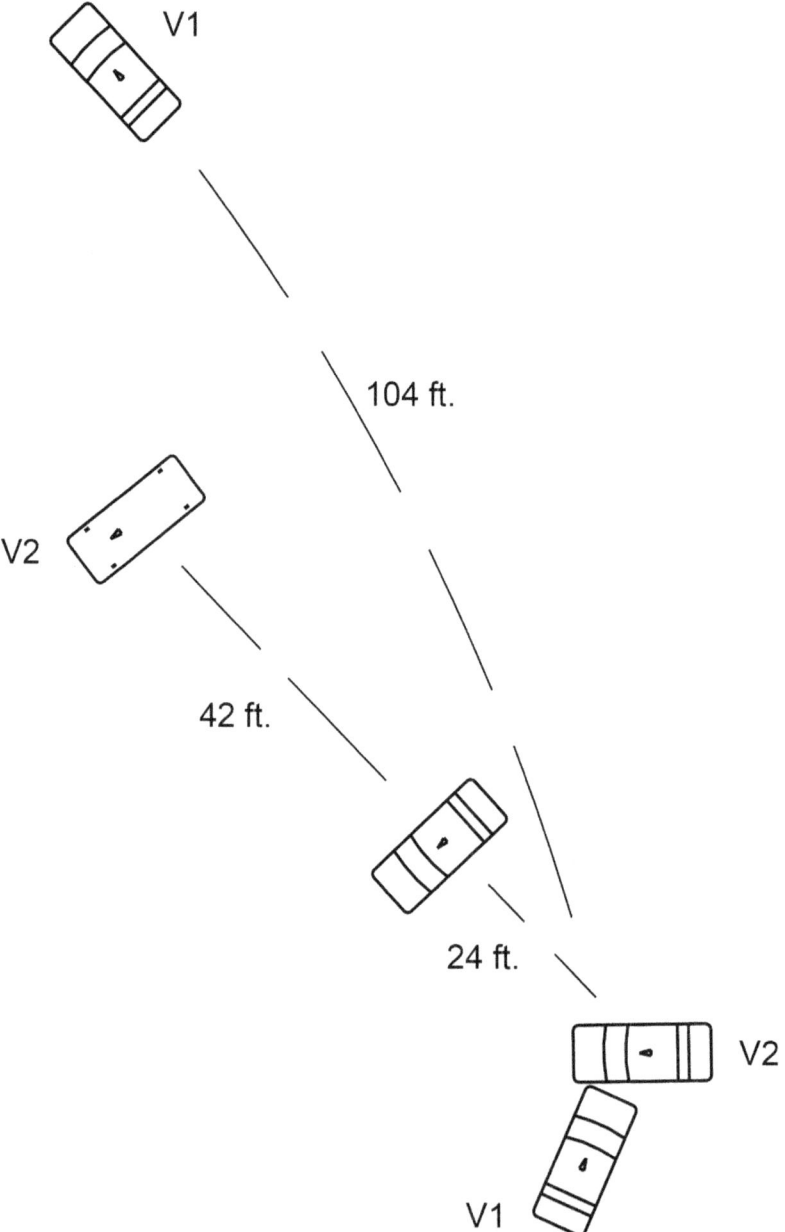

Figure 6.1. Car vs. Car Intersection Collision – Quantitative Analysis

The pavement is dry with an estimated coefficient of friction of 0.8 throughout the intersection. A firm but wet grass shoulder in the northeast quadrant has a friction factor of approximately 0.40.

Information for each vehicle is as follows:

Vehicle 1

Weight is 4200 pounds (including driver)

Damage is an angled frontal crush consistent with the impact configuration.

Left front tire is jammed

Right front tire is possibly jammed

Approach direction is N 30° W based upon the intersection geometry and the vehicle damage.

A left front tire skidmark follows a curved path that starts at N 15° E, continues for 104 feet with 40 feet on the pavement and 64 feet on the grass

A fluid (radiator) trail follows the same path just to the right of the skidmark

Vehicle 2

Weight is 3600 pounds (including occupants)

Damage to the left side includes a direct impact to the front axle

Approach direction is due east

Yaw marks indicate the vehicle's CG left the impact at N 45° E, traveled 24 feet in a counter-clockwise yaw until the right side tires struck a curb. At this point the vehicle was approximately perpendicular to its travel path. A rollover then occurred and V2 continued to roll for 42 feet, coming to rest on its roof.

An analysis of this collision will follow the previously outlined procedure.

Post-Crash

Vehicle 1: The composite coefficient of friction for the 104 feet travel path is given by

$$f_c = [0.8(40) + .40(64)]/104 = 0.55.$$

From a partial braking analysis with one front wheel braking (not shown), 66% of the weight will be on the front wheels. Therefore

$$f_e = \tfrac{1}{2}(.66)(.55) = 0.18 \text{ (Assumes RF tire is not jammed)}$$

$$S1_2 = 5.47\sqrt{(.18)(104)} = 23.7 \text{ mph}$$

$$S1_2(x) = \sin 15°(23.7) = 6.1 \text{ mph}$$

$$S1_2(y) = \cos 15°(23.7) = 22.9 \text{ mph}.$$

Vehicle 2: For the 42 feet rollover we will assume that the effective drag factor is $f_e = 0.50$. Therefore

$$S2_3 = 5.47\sqrt{(.50)(42)} = 25.1 \text{ mph}.$$

For the 24 feet yaw the average friction factor is

$$f_e = \frac{.707 + 1.00}{2}(0.80) = 0.68$$

$$\overline{S}(\text{yaw}) = 5.47\sqrt{(.68)(.24)} = 22.1 \text{ mph}.$$

Combining the two speeds yields

$$S2_2 = \sqrt{(25.1)^2 + (22.1)^2} = 33.4 \text{ mph}$$

$$S2_2(x) = \sin 45°(33.4) = 23.6 \text{ mph}$$

$$S2_2(y) = \cos 45°(33.4) = 23.6 \text{ mph}.$$

Speed Correlation

From the impact configuration shown in Figure 6.1 the following criteria must be satisfied

1
$$S2_2(y) \geq S1_2(y)$$

$$23.6 \geq 22.9$$

2
$$S2_2(x) \gg S1_2(x)$$

$$23.6 \gg 6.1.$$

Therefore, these criteria are met and no adjustments are necessary. It also is now obvious that the right front tire of V1 was not jammed and our assumption was correct.

Impact

Since a damage analysis is not going to be reliable due to the axle impact on V2, conservation of momentum is the only tool available. The expression for conservation of momentum northbound is given by

N: $\quad S1_1(y)W_1 + S2_1(y)W_2 = S1_2(y)W_1 + S2_2(y)W_2$

$$S1_1 \cos 30°(42) + 0 = 22.9(42) + 23.6(36)$$

$$S1_1(36.4) = 962 + 850 = 1812$$

158 | HIGHWAY ACCIDENTS

$$S1_1 = 49.8 \text{ mph} \approx 50 \text{ mph}$$

$$S1_1(x) = 24.9 \text{ mph}$$

$$S1_1(y) = 43.1 \text{ mph}.$$

Conservation of momentum eastbound is given by

E: $$S1_1(x)W_1 + S2_1(x)W_2 = S1_2(x)W_1 + S2_2(x)W_2$$

$$-24.9(42) + S2_1(36) = 6.1(42) + 23.6(36)$$

$$S2_1(36) = 6.1(42) + 23.6(36) + 24.9(42) = 2150$$

$$S2_1 = 59.7 \text{ mph} \approx 60 \text{ mph}$$

$$S2_1(x) = 59.7 \text{ mph}$$

$$S2_1(y) = 0 \text{ mph}.$$

Change in Speed

Changes in speed in each direction are inversely proportional to the weight of the vehicles. This ratio is equal to

$$W_1 / W_2 = 4200 / 3600 = 1.17.$$

Therefore, the change in speed ratio in each direction must be equal to

$$\Delta S_2 / \Delta S_1 = 1.17.$$

Changes in speed in the x directions are

$$\Delta S_1(x) = 24.9 + 6.1 = 31 \text{ mph}$$

$$\Delta S_2(x) = 59.7 - 23.6 = 36.1 \text{ mph}.$$

The ratio is then equal to

$$\Delta S_2(x) / \Delta S_1(x) = 36.1 / 31.0 = 1.165 = 1.17.$$

Therefore, this appears to be satisfied.

Changes in speed in the y direction are

$$\Delta S_1(y) = 43.1 - 22.9 = 20.2 \text{ mph}$$

$$\Delta S_2(y) = 23.6 - 0 = 23.6 \text{ mph}.$$

This ratio is then equal to

$$\Delta S_2(y) / \Delta S_1(y) = 23.6 / 20.2 = 1.168 = 1.17.$$

Therefore, this also appears to be satisfied.

Given the possible inaccuracies in determining the departure angles and the various assumptions regarding effective drag factors, the speeds at impact $S1_1 \approx 50$ mph and $S2_1 \approx 60$ mph are not precise values. Conducting a sensitivity analysis would most probably reveal that a ± variation of at least 5 mph would be a comfortable estimate.

Truck vs. Suv Intersection Collision

This truck/SUV intersection collision is an example that involves a somewhat different reconstruction analysis. A general layout of the scene, shown in Figure 6.2, involves one 12-foot lane northbound, one 12-foot lane southbound and four 10-foot lanes westbound. This urban intersection is surrounded by multistory buildings, including one located on the southeast corner six feet from the edges of the streets, producing a total sight distance blockage.

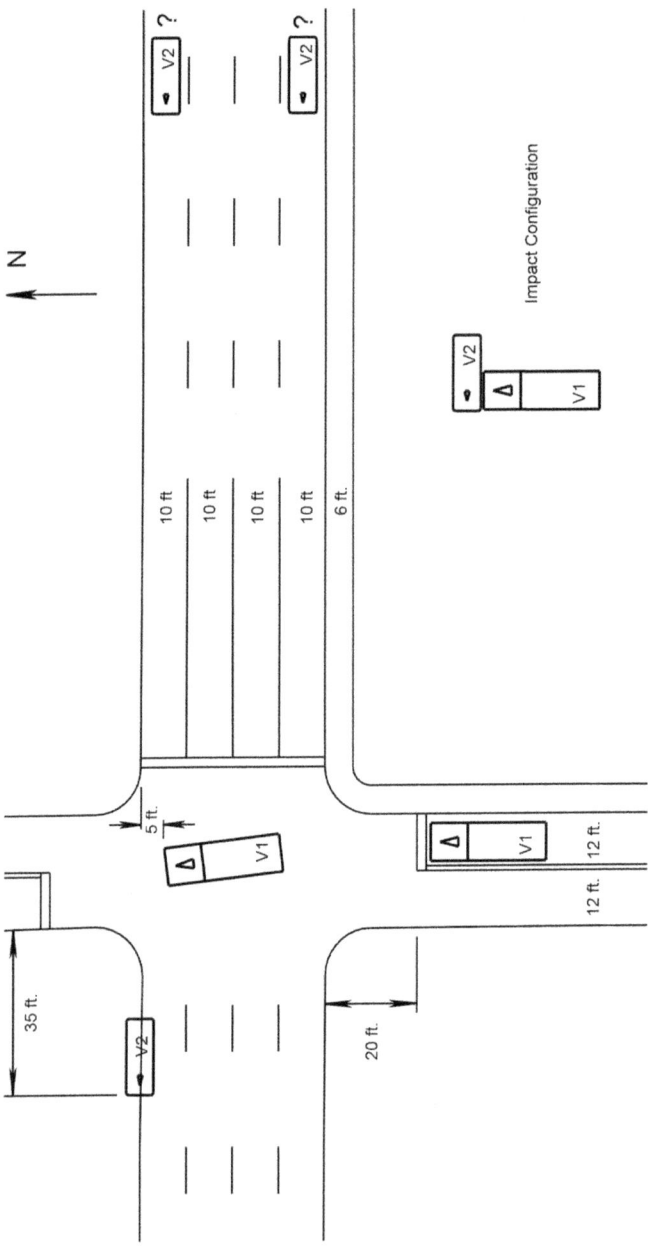

Figure 6.2 Truck vs. SUV Intersection Collision

At the time of the accident (3:00 AM) it was street-lighted and dry with a statutory speed limit of 30 mph in all directions. Both streets were travel-worn with a coefficient of friction of approximately 0.7.

Involved in the collision were a large city fire rescue truck (V1) and a private SUV (V2). City police investigating the accident interviewed the driver of the truck (V1-1), took no measurements, but did take a quality set of photographs showing each vehicle in its final rest position. No photos were taken of any street evidence that would indicate the point of impact or the post-impact vehicle trajectories. Although neither the driver of the SUV (V2-1) or the witness (W-1) were interviewed, V2-1 was cited for speeding and failing to yield to an emergency vehicle.

When a subsequent independent investigation was begun, the vehicles were no longer available and there was no physical evidence remaining at the site of the accident. Inspection of the site resulted in a scale diagram with the final rest position of each vehicle, as shown in Figure 6.2, established from the police photographs.

This photographic evidence of the vehicle damage was also sufficient to determine that the left front corner of V1 struck the left front corner of V2 at approximately a 90 degree angle as also shown in Figure 6.2. Since the front axle of V2 took the brunt of the impact there was relatively minor crush to the left side of V2. There were, in addition to this crush, heavy longitudinal scrapes down the entire left side of V2. There were also longitudinal scratches with minor damage to both the roof and the right side. There were no circular scratches to indicate any significant rotation about either the x or z axis. The post-impact motion of V2 was essentially a "barrel roll" about the longitudinal y axis.

Damage to the front of the truck was, as expected, rather small with scratch marks across the front bumper. The vehicle data of interest is:

	Weight	Width	Length	Height
V1	16,000 lbs.	8 ft.	25 ft.	Unknown
V2	3,200 lbs.	6 ft.	15 ft.	6 ft.

Interviews of both drivers and the independent witness (a county off-duty police officer) are summarized as follows:

V1-1: "I was northbound with both emergency flashers and siren operating and stopped at the stop bar as required by departmental policy. Looking left, right and left, I saw V2 very far away in the right westbound lane and, having plenty of time, I proceeded slowly across the intersection at no more than 5 mph. I was almost across when I looked again and saw V2 coming at 60 mph just before she hit me. I had no time to avoid – it was her fault."

V2-1: "I was westbound in the left lane at approximately 30 mph. The traffic light was green. I saw the truck just before it hit me. That is the last thing I remember."

W-1: "I was northbound approximately half a block behind V1. We were both going about 30 mph. Saw brake lights, heard a crash and then saw V2 airborne. I had not seen V2 before the crash and do not recall the color of the light."

Having concluded the assembly of evidence, a qualitative evaluation resulted in these initial observations:

1 From the physical evidence it is clear that V2 experienced a complete rollover and ended upright straddling the right edgeline.
2 An impact speed of 5 mph by V1 would not be sufficient to produce the rollover of V1.
3 It is obvious that V2 could not do a complete rollover and move northward only one-half of a lane width.
4 Since V1 experienced only a slight redirection and V2 basically sideswiped the front of V1, changes in speed westbound at impact would be small for both vehicles.
5 Due to this small change in speed and a relatively short post-impact slide, an impact speed of 60 mph would be totally inconsistent with the physical evidence.
6 There is agreement that V1-1 had a red light.
7 It appears that the statement of the rescue truck driver is, at best, highly inaccurate.

Post-Impact

Determination of the travel path of V2 and, therefore, the probable point of collision is the first step required for the quantitative analysis. During the rollover V2 must move laterally a distance equal to the circumference of the vehicle plus the "slippage" of 25% or more that may occur.

A minimum expected roll distance d would be equal to:

Circumference = 6 ft. + 6 ft. + 6 ft. + 6 ft. = 24 ft.
Slippage: .25(24) to 0.50(24) = 6 to 12 ft.
Total d = 30 ft. or more

Assuming that V2 is in the middle of a lane, the distance to final rest from each lane would be as follows:

Right Lane #1 d = 5 ft Not possible
            #2 d = 15 ft Not possible
            #3 d = 25 ft Possible
Left Lane   #4 d = 35 ft Possible

It is probable that V2 was, as she stated, traveling westbound in the left lane.

Due to the light damage, the effective rollover drag factor for V2 would be in the lower part of the normal range of f_e = .40 to 0.60. Using f_e = 0.45 would appear to be reasonable. We will also assume d = 35 feet. The post-impact speed of V2 in the y direction (north) is then equal to

$$S2_2(y) = 5.47\sqrt{(.45)(35)} = 21.7 \text{ mph}.$$

Referring to Figure 6.2, V2 would have traveled a distance west equal to:

$$d = 2 \text{ ft (to CL)} + 12 \text{ ft (lane width)} + 35 \text{ ft} = 49 \text{ ft}.$$

An effective drag factor of .30 to 0.40 would be appropriate for this sliding action. We will assume f_e = 0.35. The post-impact speed of V2 in the x direction (west) is equal to

$$S2_2(x) = 5.47\sqrt{(.35)(49)} = 22.6 \text{ mph}.$$

Since V1 and V2 maintained contact as V2 slid across the front of V1, their post-impact speeds northbound were essentially equal. That is

$$S1_2(y) = S2_2(y) = 21.7 \text{ mph}.$$

A cross check on this post-impact speed of V1 is to determine if V1 could have stopped at its point of final rest. The distance traveled would be equal to

$$d = 8 \text{ ft (2 ft into lane \#4)} + 2 \text{ (10 ft. lanes)} + 5 \text{ ft} = 33 \text{ ft}.$$

The required drag factor to stop in this distance is

$$f_e = \frac{S^2}{30d}$$

$$f_e = \frac{(21.7)^2}{30(33)} = 0.48.$$

This represents a hard braking consistent with the situation.

At this stage it is appropriate to consider the possibility that the impact occurred in Lane #3, redoing the calculation with the roll distance of V2 equal to 25 feet and the braking distance of V1 equal to 23 feet.

$$S1_2(y) = S2_2(y) = 5.47\sqrt{(.45)(25)} = 18.3 \text{ mph}$$

$$f_e = \frac{(18.3)^2}{30(23)} = 0.49.$$

Therefore, this is also consistent with the impact occurring in Lane #3.

Impact

From conservation of momentum the changes in speed during impact are inversely proportional to their weights. Therefore

$$\Delta S_1(y) = \Delta S_2(y) \frac{W_2}{W_1}.$$

Since $\Delta S_2(y) = S2_2(y) = 21.7$ mph and the weights are known, the longitudinal change in speed of V1 is equal to

$$\Delta S_2(y) = 21.7 \left(\frac{3200}{16000} \right)$$

or

$$\Delta S_2(y) = 4.3 \text{ mph}.$$

The impact speed of V1 would then be equal to

$$S1_1 = 21.7 + 4.3 = 26 \text{ mph}.$$

An estimate of the speed of V2 at impact is not subject to a reliable calculation. Since the loss in speed during impact is only due to friction as it slides across the front of V1, all we can conclude is that this value will be small. This impact speed of V2 may be expressed as:

$$S2_1 = S2_2(x) + \Delta S_2(x)$$

or

$$S2_1 = 22.6 \text{ mph} + \text{some small unknown value}.$$

There is, therefore, no creditable evidence that V2 was exceeding the speed limit of 30 mph.

Pre-Crash

V1-1 said he stopped at the stop bar. In order to reach a speed of 26 mph (38 ft/sec) over a distance equal to

$$d = 20 \text{ ft (to lateral line)} + 2 \text{ ft (into Lane \#4)} = 22 \text{ ft}.$$

would require an acceleration time of

$$t = \frac{2d}{V} = \frac{2(22)}{38} = 1.16 \text{ sec}$$

and a corresponding acceleration rate of

$$a = \frac{V}{t} = \frac{38}{1.16} = 32.8 \text{ ft/sec}^2$$

that is, of course, not possible.

If we assume that the impact occurred in Lane #3, we can use the recalculated post-impact speed of V1

$$S1_2(y) = 18.3 \text{ mph}$$

and calculate the impact speed of V1 as equal to

$$S1_1 = 18.3 + 18.3 \left(\frac{3200}{16000} \right)$$

or

$$S1_1 = 18.3 + 3.7 = 22 \text{ mph (32 ft/sec)}.$$

The time required to travel a distance

$$d = 22 \text{ ft} + 10 \text{ ft} = 32 \text{ ft}$$

will be equal to

$$t = \frac{2d}{V} = \frac{2(32)}{32} = 2 \text{ sec}$$

and the required acceleration rate would be equal to

$$a = \frac{V}{t} = \frac{32}{2} = 16.0 \text{ ft/sec}^2$$

This truck would be unable to achieve that rate of acceleration.

If V1 had stopped closer to the lateral line, the acceleration rate required would have been even greater. Therefore, it is safe to conclude that V2 was traveling in the left westbound lane and that V1 did not stop for the red light.

Avoidance

An important factor in this type of accident, and in this type of environment, is that it is quite likely that the driver of the SUV (V2-1) did not hear the siren until just before impact. A number of research studies have shown that buildings can effectively block the sound of a siren from intersecting streets. It has also been shown that, even without intervening structures, a driver in a closed passenger vehicle may not hear the siren until it is very close. A further finding is that, even when you finally hear the siren, you generally have no idea where it is located.

Since V1 did not stop and entered the intersection well in excess of 20 mph and V2 was traveling at approximately 30 mph, neither driver would have had sufficient time to execute any meaningful avoidance maneuvers. They would not see each other until less than one second prior to impact. This would be barely enough time to react, let alone make any significant reduction in speed.

Causation

There is no evidence that V2-1 should have been able to avoid this collision. V1-1 entered a blind intersection without stopping for the red light, a reckless action that was the obvious cause of the accident.

Car vs. Car Intersection Collision – Qualitative Analysis

The following example illustrates the process for an analysis that yields only a qualitative reconstruction and a limited causal analysis. This accident that occurred late at night involved a police patrol car V1 and a private passenger car V2. First responders (patrol officers and fire rescue) stabilized the scene and transported both drivers to a local hospital. A police accident investigator was also called to the scene, interviewed the one independent witness, and photographed the vehicles at final rest. No measurements were taken of the vehicle damage, the vehicle final rest positions, or the location of all the other physical evidence. A police report was issued based upon the investigator's observations and, in accordance with the statement of the independent witness W1, that V1 was eastbound, turned left, and struck V2 that was traveling westbound with a flashing yellow light.

A subsequent investigation was conducted after both vehicles had been destroyed and all physical evidence at the scene had disappeared. County records revealed that at the time of the accident flashing red signals controlled north-south traffic, and east and westbound vehicles would have had a flashing amber signal. This creates somewhat of a problem for north and southbound vehicles since, after stopping, they must cross seven travel lanes and a median without any assigned right of way.

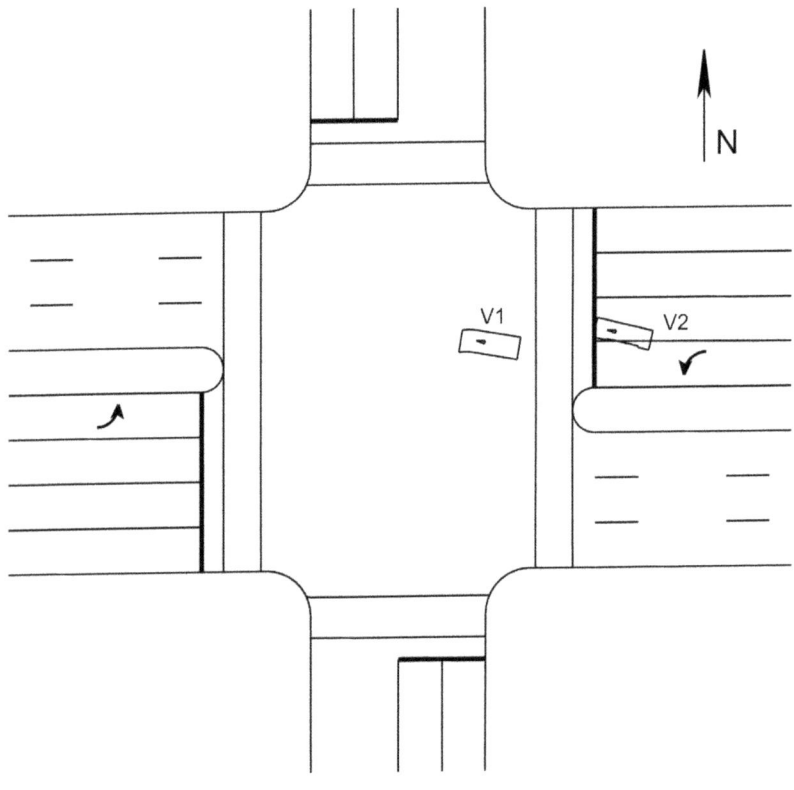

Figure 6.3. Car vs. Car Intersection Collision – Qualitative Analysis

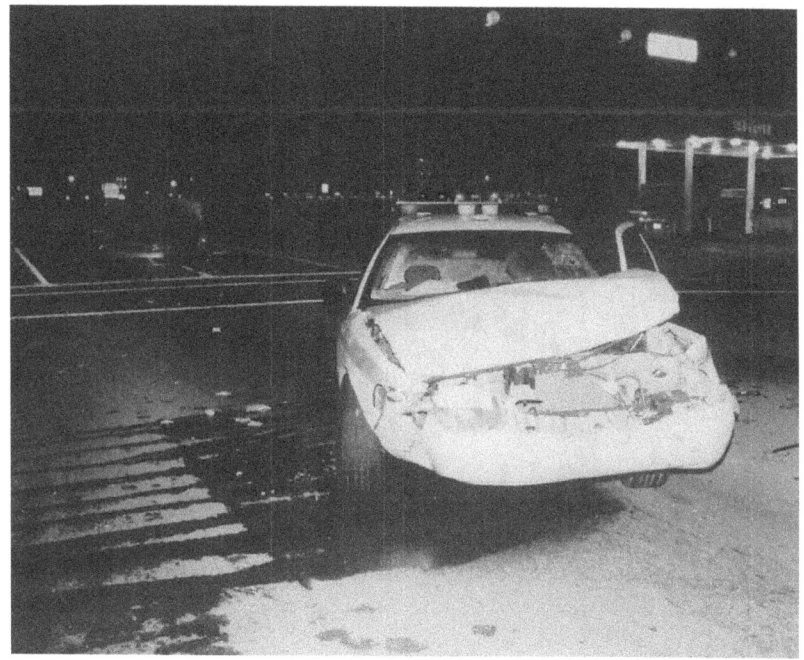

Figure 6.4. V1 at Rest

Figure 6.5. V2 at Rest

170 | HIGHWAY ACCIDENTS

Figure 6.6. V1 Fluid Trail

Using the available police photographs at the site of the accident, the approximate final rest position of each vehicle was determined and then placed upon a diagram of the scene as shown in Figure 6.3. A frontal view of V1, shown in Figure 6.4, indicates its position at rest as well as the damage pattern. Figure 6.5 indicates the final rest position of V2 and the damage to its left side (somewhat obscured by a detached door). A second view of V1, shown in Figure 6.6, also indicates its final rest location and the presence of a fluid trail and a smeared tire mark.

Both drivers, V1-1 and V2-1, along with the independent witness W1 were interviewed and gave the following descriptions of the accident events. V1-1 stated she was eastbound in the inside through lane when V2 entered the intersection from her right (traveling northbound). She attempted to brake and swerve left to avoid but struck V2 in the left side. V2-1 stated he was westbound when the police car (V1) came out of nowhere and hit him, knocking him momentarily unconscious.

The first step in the qualitative reconstruction process is the determination of the relative orientation and movement of the two vehicles at impact. Front damage to V1 (see Figure 6.4) and damage to the left side of V2 (see Figure

6.5) clearly indicate a broadside, approximately perpendicular, impact with V1 possible rotated slightly counter-clockwise from 90 degrees. Displacement to left of the front of V1 clearly indicates that V2 was moving from right to left across the front of V1. A snagging action is apparent by the extent of this sidesway and by the indentation on the right side of the front bumper of V1 that matches the damage to V2 just forward of the left rear wheel. This damage would have occurred as the front of V1 penetrated the left side of V2. This impact configuration shown in Figure 6.7 does not, however, establish the actual travel directions of the vehicles or their locations at impact.

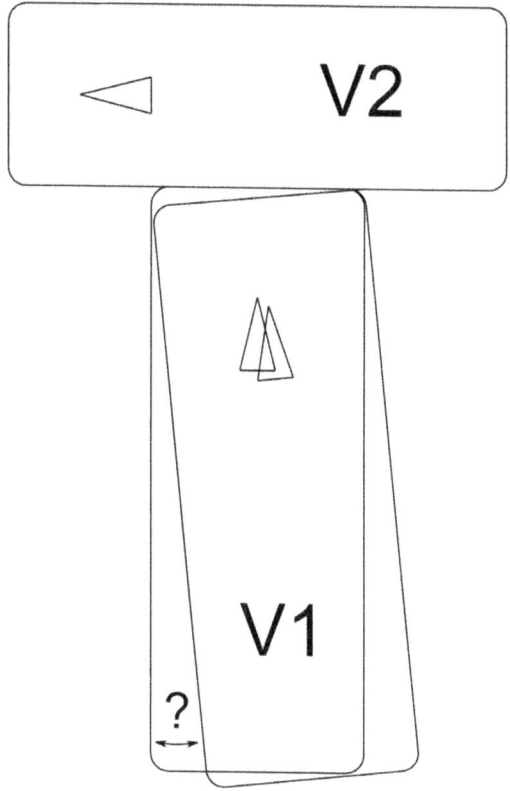

Figure 6.7. Impact Configuration

A determination of the initiation of the post-impact trajectories is the next step in the process. V2 must have been slowed, pushed to the right and caused to begin a counter-clockwise rotation. V1 must be decelerated, redirected to the left and rotated counter-clockwise. Referring to Figure 6.6, the fluid trail (probably from the radiator) also indicates this redirection and the smeared tiremark is consistent with a right rear tire side scuff due to the rotation of V1.

Movement from impact to final rest of V1 can be inferred from the fluid trail that appears to lead to its final location. There is no apparent physical evidence showing V2's post-impact trajectory but its final rest location, as we will see shortly, is consistent with the evidence.

The final step in this analysis is to determine the approximate area of the collision and the vehicle orientations at impact. A careful examination of Figure 6.6 also shows the rest position of V1 and the end of the median, thus establishing that the impact occurred south of the median and that V1 must have been traveling eastbound at impact. The previously determined impact configuration clearly establishes that V2 must, therefore, have been traveling northbound.

Conservation of momentum must also be satisfied. Post-impact movement of the vehicles was northeast. Therefore, the combined pre-crash momentum of both vehicles must also be northeast. This is, therefore, consistent with V1 traveling eastbound and V2 approaching northbound.

An evaluation of the drivers' and witnesses' statements indicates that the only accurate account of the accident events was given by V1-1, the driver of the police vehicle. The other driver V2-1 may have been stunned and believed, from his final rest position and orientation, that he had in fact been traveling westbound and had never seen V1. It is, of course, also quite possible his version of the accident was simply dishonest and self-serving. It is probable that the witness saw only the tail end of the accident sequence or just the final rest configuration from which he assumed what must have happened. This is a common occurrence that demonstrates the lack of reliability of witness testimony.

In the collision analysis it is clear what happened, but it is unlikely that a reasonable estimate of speeds can be determined. The precise direction of V1's approach is unclear due to the possibility of a pre-crash swerve. The actual location of the vehicles at impact cannot be determined. It is also not possible to make any reliable estimate of the departure angle of either vehicle.

A lack of significant frontal crush to V1 and the limited penetration to the side of V2 indicates that V1 was probably not exceeding the speed limit

(40 mph) at impact. There is, however, no physical evidence regarding any pre-crash braking. Therefore, the pre-crash speed of V1 is unknown. The post-impact speed and the change of speed at impact of V2 are also unknown. Therefore, we cannot determine if it was possible for V2-2 to have stopped prior to entering the intersection. Even if it were possible, we still would not know if in fact he did stop.

A complete causal analysis is not possible since we cannot say for certain that V1-1 was not speeding nor can we be certain if her speed contributed in any way to the causation of the accident. As previously stated, we do not know if V2-1 stopped for the flashing red signal, but if V1-1 was not speeding, then clearly he caused the collision by entering the intersection without yielding the right of way.

Roadside Collisions

A roadway departure is usually the precipitating event in a single vehicle accident, but also may occur as the result of a collision with another vehicle. This can result in a rollover or an impact with some roadside object. If the shoulder and/or roadside is relative level, stable and open, the driver may have sufficient time and space to safely recover.

Rollover

This recovery is often prevented by an unstable surface such as soft ground. Turning to regain the roadway can cause the outside tires to furrow into the soil and trip the vehicle into a rollover. This rollover can be analyzed as shown in Chapter 3 Vehicle Dynamics, Rollover.

An attempt to recover by turning on a slope, as shown in Figure 6.8, can also precipitate a rollover. Determination of the speed S required to produce the rollover can be done in essentially the same manner as shown in Chapter 5 Analysis Procedures, Critical Speed. At impending tipping, the vehicle will pivot about the outside edge of the downhill tires.

174 | HIGHWAY ACCIDENTS

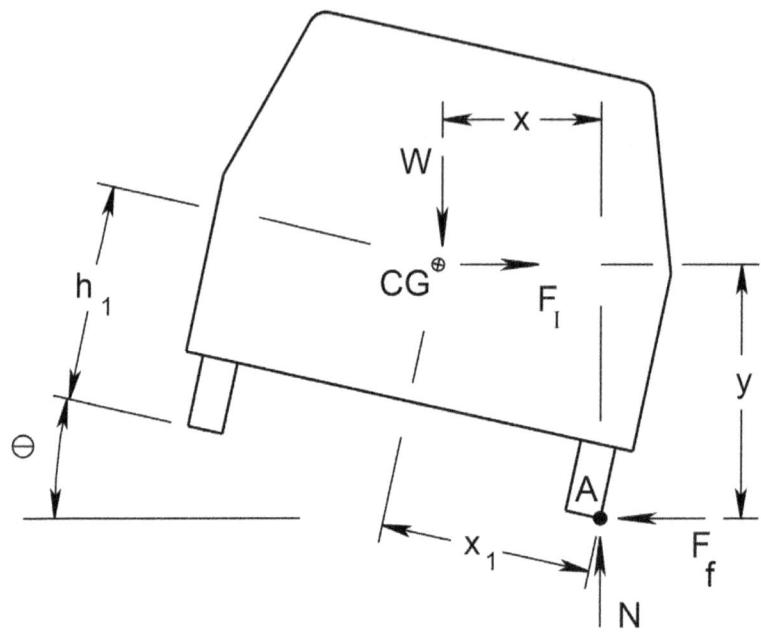

Figure 6.8. Rollover on Slope

Summing moments about Point A yields

$$F_I[y] - W[x] = 0 \tag{6.1}$$

where

F_I = centrifugal force.

$$F_I = \frac{W}{g}\frac{V^2}{R}$$

where

R = radius of the vehicle path
W = weight of the vehicle
$y = x_1 \sin\theta + h_1 \cos\theta$
$x = x_1 \cos\theta$
θ = slope of the ground
x_1 = distance from the CG to edge of the tire
h_1 = height of the CG above ground.

Substituting into Equation (6.1) yields

$$\frac{W}{g}\frac{V^2}{R}(x_1 \sin\theta + h_1 \cos\theta) = (Wx_1 \cos\theta).$$

Eliminating W and dividing by $x_1 \cos\theta$ yields

$$\frac{V^2}{gR}\left(\tan\theta + \frac{h_1}{x_1}\right) = 1.$$

Since $\tan\theta = $ slope e in percent

$$V^2 = \frac{gR}{(e + h_1/x_1)}.$$

Expressed in terms of speed

$$(1.475)^2 = \frac{32.2R}{(e + h_1/x_1)}$$

or

$$S^2 = \frac{15R}{(e + h_1/x_1)}$$

or

$$S = 3.87\sqrt{R/(e + h_1/x_1)}. \tag{6.2}$$

As a cross check, when $e = \tan\theta = 0$, this reduces to

$$S = 3.87\sqrt{Rx_1/h_1}$$

that is the previously derived equation for the rollover critical speed.

Using Equation (6.2) to estimate the vehicle speed at rollover may *not* be valid due to the following potential limitations:

1 The height h_1 of the CG may be unknown.
2 The appropriate adjustments to x_1 and h_1 are often difficult to determine.
3 The path of the vehicle is not always visible.

4 The path radius R may be rapidly changing and, therefore, difficult to measure.

5 An undetectable sudden steering input may precipitate the rollover.

As previously noted, the determination of the speed at rollover can also be estimated using the methodology described in Chapter 3 Vehicle Dynamics, Rollover.

Rigid Objects

Although rollovers generally have very serious consequences, impacts with rigid roadside objects produce even more fatalities and severe injuries. Common median and roadside hazards include the following:

- Large trees
- Utility and sign poles
- Underpass/overpass and bridge support structures
- Culverts and other drainage structures
- Terminations of longitudinal barriers
- Construction equipment and material

Impacts with these roadside hazards are often more severe than vehicle-to-vehicle impacts since the resultant changes in speed (ΔV's) are generally greater. For example, if a vehicle V1 traveling at 60 mph rear ends a similar stopped vehicle V2, each will experience a change in speed of 30 mph. If, however, V1 strikes a rigid object at 60 mph, the change in speed will be 60 mph. This reminds us that *change* in speed is the real measure of crash severity that directly relates to injury severity. The calculation of the impact speed *FBE* with a rigid object is covered in Chapter 4 Impact Dynamics.

Roadside Protective Devices

Roadway departures can also result in collisions with protective devices that are utilized to alleviate the consequences of a roadway departure by containing the vehicle or by reducing the severity of an impact. These devices consist of three basic types:

1 Redirection devices such as rigid concrete barriers, flexible guard rails, and tensioned cable barriers

2 Energy absorbing (EA) devices

3 Breakaway light poles or sign supports

Redirection Devices

Containment of a errant vehicle is accomplished by placing a longitudinal barrier on the outside of the shoulder. The purpose of the barrier is to prevent the vehicle from entering a hazardous roadside or crossing the median into on-coming traffic by redirecting the vehicle parallel to the roadway. A median barrier that failed due to lack of steel reinforcing bars and quality concrete is shown in Figure 6.9.

Figure 6.9. Failed Median Barrier

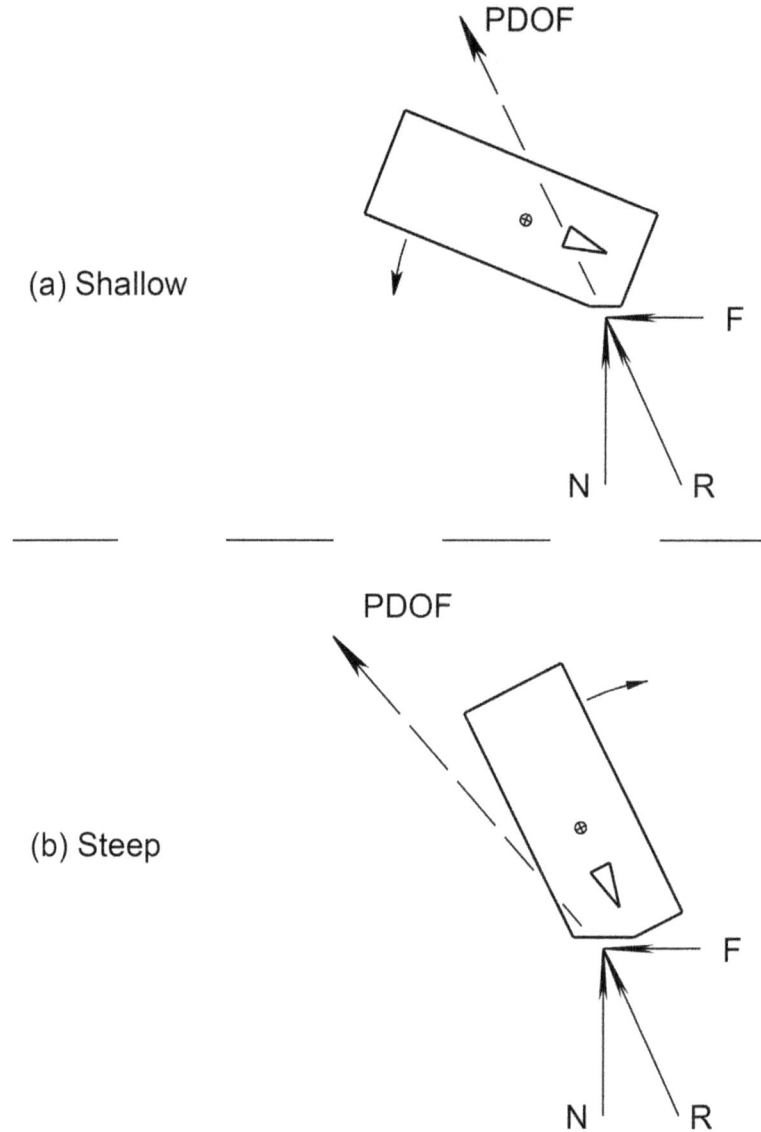

Figure 6.10. Rigid Barrier Impact

As shown in Figure 6.10(a), an impact with a rigid longitudinal barrier at some shallow angle will produce a normal force N and a friction force F. Since the direction (PDOF) of the resultant force R passes in front of the CG, the vehicle will rotate counter-clockwise as intended and be redirected into a path approximately parallel to the roadway. This impact configuration will produce lateral and longitudinal changes in speed that generally will not cause serious injuries. Following the initial impact the rotation frequently causes a side slap to the right side of the vehicle as well as a subsequent sideswiping action. The speed loss during this impact is generally small and difficult to estimate.

At steep impact angles, as shown in Figure 6.10(b), the resultant force will be directed to the other side of the CG, thus producing a clockwise rotation. This may or may not result in a subsequent impact to the left rear of the vehicle. In this impact configuration the resultant force on the vehicle is primarily longitudinal and, therefore, has the potential for significant crush, a large longitudinal change in speed and serious injuries. A reasonable estimate of the impact speed can generally be achieved following the methodology described in Chapter 4 Impact Dynamics.

An example of the frontal crush to a vehicle that struck a concrete barrier at a high angle is shown in Figure 6.11. In this case there was no significant rotation, but there was some sliding along the barrier, as indicated by the horizontal scratches across the front of the vehicle.

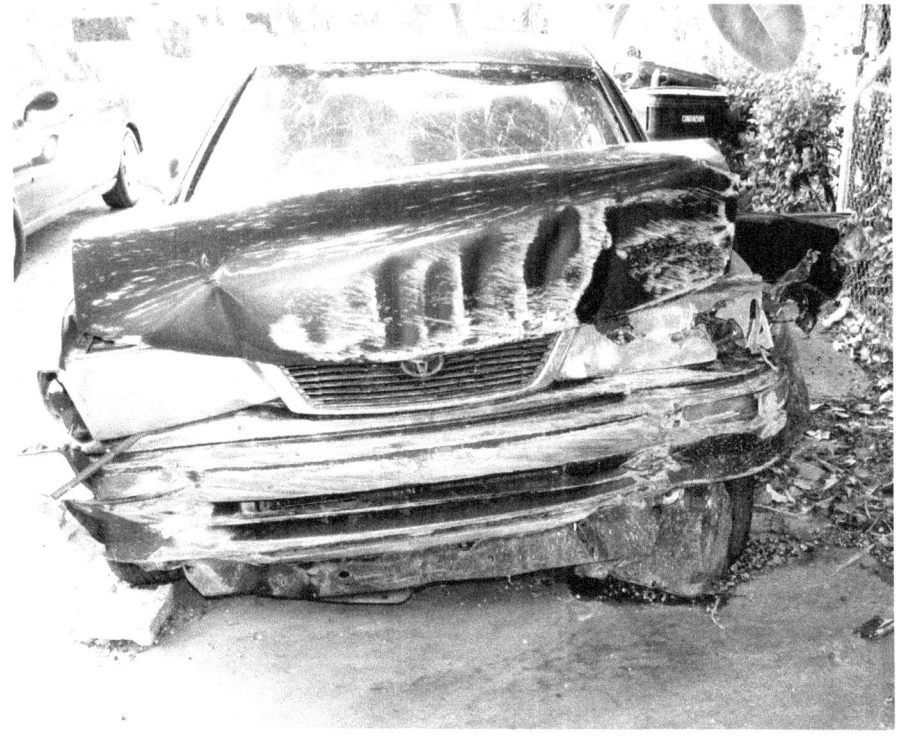

Figure 6.11. Frontal Impact

The linear crush across the 60-inch wide front varied from 12 inches on the right side to 18 inches in on the left side. Using the average crush 15 inches and the stiffness coefficients $b_0 = 5.0$ and $b_1 = 1.5$ obtained from crash tests, the impact speed S would be given by

$$S = 5.0 + 1.5(15) = 27.5 \text{mph}.$$

A reasonable speed estimate would be 25 to 30 mph. The angle θ of impact can be determined as follows

$$\tan \theta = (18 - 12) / 60 = 0.10$$

or

$$\theta = 5.7 \text{ degrees from perpendicular.}$$

Collisions with flexible barriers such as guardrails result in motions similar to rigid barriers. Determination of the impact speed of high angle impacts is, however, more complicated. Using vehicle crush to estimate the speed will yield an estimate that is lower than the actual speed since some kinetic energy is also lost due to deformation of the barrier. This loss is generally difficult to estimate, but a lower boundary for the range of impact speed can be established.

A vehicle striking a cable barrier usually produces a large deflection and increase in the cable tension. This type of crash is rarely a candidate for a quantitative reconstruction.

Energy Absorbing Devices (EA)

Energy absorbing devices are designed to "soften" impacts with rigid objects rather than trying to avoid them. A typical application would be placing a collapsible EA device at the termination of a rigid barrier. A vehicle striking this head-on will suffer frontal crush and will deform the device as shown in Figure 6.12.

Figure 6.12. Energy Absorbing Device

The kinetic energy of the vehicle will be dissipated through the work done crushing the vehicle and deforming the EA device. If the crush coefficients for the vehicle and the stiffness characteristics of the EA device are known, then the total work done, the total energy loss, and the impact speed can be calculated.

Determination of the energy loss from vehicle crush can be achieved by using the methodology described in Chapter 4 Impact Dynamics. The fixed barrier equivalent speed (FBE) S determined from the vehicle crush is

$$S = b_1 c \text{ or } S = b_0 + b_1 c.$$

The velocity would then be equal to

$$V = (1.47 S)$$

and

$$V^2 = 2.16 S^2$$

since kinetic energy is

$$KE = \frac{W}{2g}V^2.$$

The kinetic energy loss KE_V due to the crush of the vehicle is equal to

$$KE_V = \frac{W}{2(32.2)}(1.47S)^2$$

or

$$KE_V = \frac{W}{30}S^2. \tag{6.3}$$

Since the Energy Absorbing Device is essentially a plastic spring, the $Work_{EA}$ done on the device is

$$Work_{EA} = \frac{1}{2}kc_{EA}^2$$

where

$k =$ stiffness coefficient of the device

$c_{EA} =$ deformation of the device.

Since the kinetic energy KE_{EA} loss in deforming the device is equal to the work done

$$KE_{EA} = \frac{1}{2}kc_{EA}^2 \tag{6.4}$$

and, therefore, the total energy loss is equal to the sum of equations (6.3) and (6.4)

$$KE_T = \frac{W}{30}S^2 + \frac{1}{2}kc_{EA}^2.$$

The vehicle speed can then be found as follows

$$\frac{W}{2g}V^2 = KE_T$$

or

$$V = \sqrt{\frac{2g}{W} KE_T}$$

or

$$V = 8\sqrt{KE_T / W}$$

and

$$S = 5.47\sqrt{KE_T / W}.$$

The work done by the force acting upon the car is equal to the total kinetic energy, but the distance through which the force acts will be the sum of vehicle crush and the deformation of the device. Therefore, the force required and the corresponding deceleration will be significantly less than an impact to a rigid barrier termination. This will, of course, result in less vehicle damage and reduced injury severity.

Breakaway Supports

Supports for signs are a necessary evil for the safe and efficient operation of our streets and highways. Most signs are placed on lightweight posts that break or bend over when struck by a car, thus reducing them to a minor hazard. Large overhead signs on freeways can generally have their supports placed behind barriers, thus creating a moderate as opposed to a severe problem.

For many years roadway lighting was considered an economically affordable benefit for the convenience and safety of the motorist. These lights were (and some still are) supported by rigid wooden, concrete or steel poles. The resulting hazard to errant vehicles significantly offsets the safety benefit of roadway lighting. An important improvement in safety has been the development and the widespread use of breakaway light poles for high-speed highways. They are normally tapered tubing made of a strong, lightweight aluminum alloy. The base is generally a strong, but brittle, aluminum casting that can withstand wind loads and minor impacts. However, when struck at highway speeds, this frangible base will shatter and the pole will be pushed away.

184 | HIGHWAY ACCIDENTS

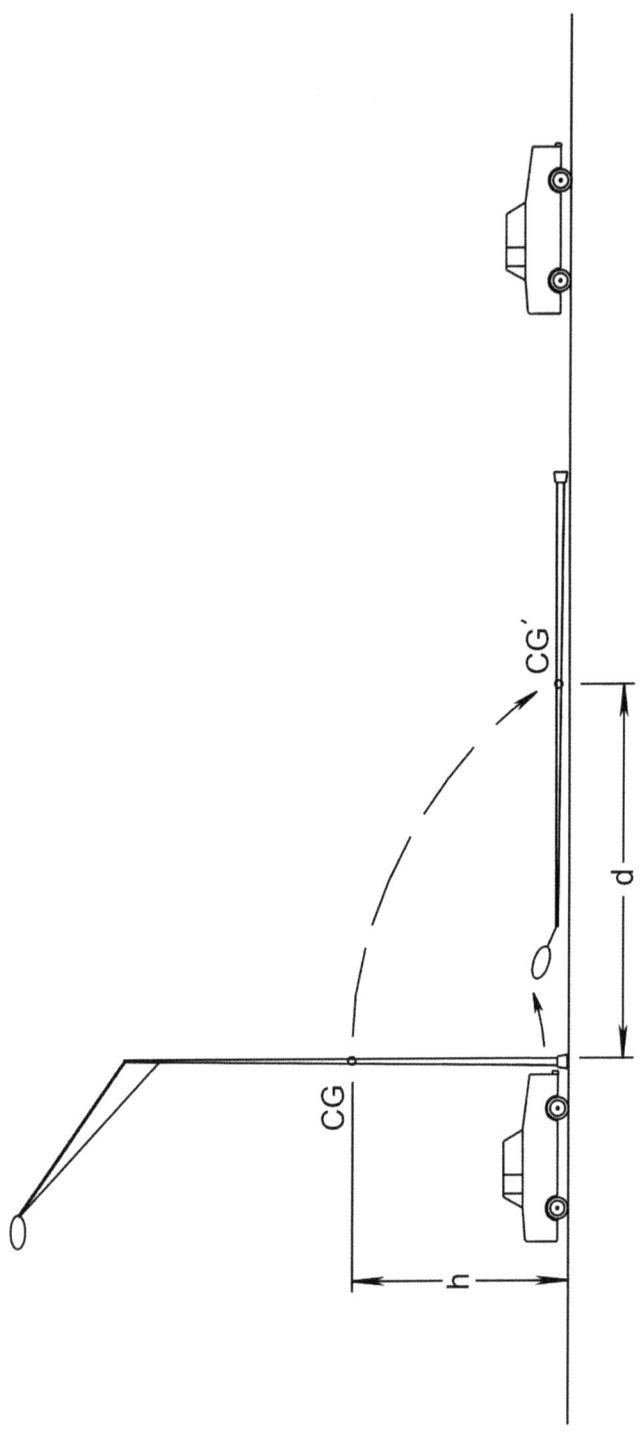

Figure 6.13. Breakaway Pole

Extensive testing programs and the investigation of many accidents reveal that the breakaway pole impact generally occurs as shown in Figure 6.13. Upon impact the base fractures and the bottom of the pole is pushed forward and the pole is rotated. The center of gravity of the pole will drop a distance h and travel forward a distance d. Meanwhile the car will pass under the pole and continue forward at a reduced speed. To illustrate the dynamics of the event, let us look at a hypothetical example. Assuming that the weight of the car is 3000 pounds and the weight of the pole is 250 pounds, and an impact speed of 15 mph will just fracture the base. At an impact speed of slightly less than or equal to 15 mph the car will stop at impact with a ΔV of 15 mph. This can produce injuries, particularly for unrestrained occupants. Impacting the pole at slightly greater than 15 mph will also produce a ΔV of approximately 15 mph and the pole will fall on top of the car, causing additional damage.

Assume the car is traveling 70 mph at impact and pushes the CG of the pole forward a distance of d = 25 feet while dropping a distance d = 16 feet. The severity of this collision can be expressed as the resultant ΔV. Note that if this were a rigid pole the ΔV will be equal to 70 mph — almost certainly a fatal accident.

It is convenient to consider the impact in two distinct phases. The first phase, breaking the base, dissipates energy through crush to the front of the car and fracturing of the base of the pole. This loss in kinetic energy is equal to that lost when the vehicle strikes the pole at 15 mph. Using our abbreviated format this may be expressed as follows

$$S_2^2 = S_1^2 - \overline{S}^2$$

where

S_1 = impact speed
S_2 = post-fracture speed
$\overline{S}$ = speed required to break the base.

Substituting for S_1 and $\overline{S}$ yields

$$S_2^2 = (70)^2 - (15)^2$$

or

$$S_2 = 68.4 \text{ mph}$$

or the initial ΔV is equal to

$$\Delta V_1 = 1.6 \text{ mph.}$$

Note that the damage to the front of the vehicle will be essentially the same as that produced by a 15-mph impact with a rigid pole.

The second phase, the pole acceleration, is solved by using projectile motion to find the speed of the pole and then conservation of momentum to find the final post-impact speed of the car. Since the pole's CG drops a distance $d = 16$ feet, the time required to fall to the ground would be given by

$$t = \sqrt{2d/g}$$

$$t = \sqrt{2(16)/32}$$

$$t = 1.0 \text{ sec.}$$

Since the horizontal speed of the center of gravity of the airborne pole is constant, the post-impact speed of the pole is given by

$$V_p = \frac{d}{t} = \frac{25}{1.0} = 25 \text{ ft/sec}$$

or

$$S_p = 17.0 \text{ mph.}$$

Conservation of momentum dictates that the change in speeds (ΔV's) of the two objects is inversely proportional to their weights. That is

$$\frac{\Delta V_c}{\Delta V_p} = \frac{W_p}{W_c}$$

or

$$\Delta V_c = \frac{W_p}{W_c} \Delta V_p.$$

Substituting yields

$$\Delta V_c = \frac{250}{3000} \quad (17)$$

$$\Delta V_c = 1.4 \text{ mph}.$$

and, therefore, the final post-impact speed S_3 of the car is equal to

$$S_3 = S_2 - \Delta V_c$$

$$S_3 = 68.4 - 1.4 = 67 \text{ mph}.$$

This yields a total change in speed of approximately 3 mph and probably no resultant injuries for this impact with a breakaway light pole.

Without showing the calculations, an impact speed of approximately 30 mph will produce a speed change of approximately 5 mph. If the pole, for example, weighed 500 pounds, the increase in ΔV would still be only 1-2 mph.

An attempt to reconstruct the speed at impact for this type of accident is going to be difficult. The weight and center of gravity of the pole may not be easy to obtain. Accurate estimates of h and d depend upon how the pole strikes the ground and how far it slides along the ground after landing. A good estimate of the kinetic energy required to fracture the base of the pole may not be available. If, however, the post-impact speed can be determined, adding 3-7 mph to that value would probably yield a reasonable estimate. Again, the inability to reconstruct this type of accident may be frustrating to the analyst, but the breakaway pole has saved a huge number of lives and has resulted in many fewer serious injuries.

One last thought. If you are traveling at highway speed and are about to hit the pole, braking will actually increase the severity of the impact. However, if you are not going to brake, make sure it is in fact a breakaway pole.

Pavement/Shoulder Drop-off

An inadvertent shallow angle departure from the travel lane to the shoulder is a common occurrence that usually has no adverse consequences. An easy recovery and return to the roadway can be accomplished if the shoulder is constructed of a stable material and if there is a smooth transition between the two surfaces.

Pavement/shoulder drop-offs occur all too frequently and present a serious roadside hazard. These are often caused by poorly constructed shoulders and/or an inadequate inspection/maintenance program. Drop-offs are also created during repaving, widening and other construction activity. They are a frequent causal factor initiating unintended maneuvers in single vehicle accidents.

In the classic pavement/shoulder drop-off accident, loss of control begins when the right side tires suddenly fall off the edge of the pavement. Alerted, the driver will immediately attempt to turn left and return to the travel lane. The success or failure of this maneuver depends upon the type of vehicle, the tire contour, and the depth and shape of the discontinuity.

For a motorcycle this event almost always results in a disastrous rollover. For a normal passenger car this becomes a problem when the vertical face of the drop-off is greater than two or three inches. As shown in Figure 6.14, when the depth h_1 of drop-off becomes greater than the height h_2 of the "bulge" in the tire sidewall, the tire will not tend to climb over the lip of the pavement.

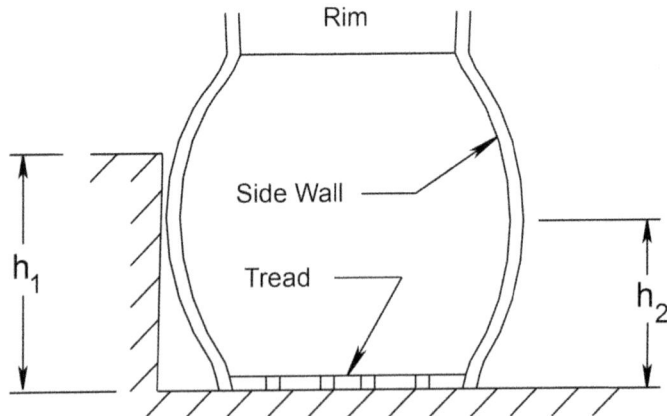

Figure 6.14. Pavement – Shoulder Drop Off

An appropriate response to this situation is to slow down until recovery is possible or to move further to the right and return at a steeper, more manageable angle. Unfortunately, the reaction to the sudden jolt and the necessity to regain the travel lane will usually prompt an immediate left-steering input that does not produce any change in direction.

Additional steering input takes up any slack in the steering linkage and begins to twist the tire. If the drop-off is not too deep, continued turning of the steering wheel will eventually produce a left turn. Similar to releasing a rubber band, the right front tire mounts the pavement lip and reenters the travel lane. This reentry will, however, be at an angle much greater than expected, causing an undesired sharp swerve to the left. This may cause an excursion into the opposing travel lane and a roadway departure to the left or a counter steer to the right. This steering effort is frequently an over-correction, leading to a second right side departure and a possible off-road impact or rollover. Like most emergency situations, the drop-off is unexpected, and the proper solution is unknown and has not been carefully thought out prior to its occurrence. Unlike airline pilots, automobile drivers are not required to practice recovery from emergency situations.

In the investigation of a single vehicle roadway departure or a crossover head-on collision an investigation of the roadside upstream from the crash event is a must. If a drop-off is found, a careful examination may show a tire track and/or scuff marks or rim marks at the reentry point. An examination of the right front wheel may reveal scrub marks on the inside sidewall and /or scraping of the rim.

Reconstruction Technology 7

Techniques for the analysis of physical behavior and for making engineering decisions have evolved slowly and erratically. Although the universally used trial and error method resulted in many great failures, it did lead to successes such as the Roman arch and the Egyptian pyramid. Developments in mathematics such as the Arabic invention of the zero, and the replacement of Roman numerals, allowed for a more rational quantification of physical phenomena. Parallel scientific efforts in the field of physics included the pioneering advances in particle motion by Galileo and Newton. Frustrated by the limitations of geometry in his attempt to understand and describe motion, Newton invented calculus. This then allowed for a proper description of the relationships among time, distance, speed, acceleration, force, mass, work, energy, impulse, and momentum.

Using the concept of the logarithm, the slide rule allowed for a rapid solution of many mathematical computations. While it had limitations on accuracy it was used to design buildings, bridges, ships, aircraft and many other complex structures and machinery, most of which worked as intended. The next related innovation was the hand calculator that permitted a much broader range of mathematical computations with significantly greater accuracy. This invention was instrumental in the transition from the analog to the digital age. There is, however, one fundamental drawback to using digital calculations –they are contrary to our natural analog thinking process. Finally, the computer has revolutionized virtually all engineering and technological analyses, including accident reconstruction.

Computer-Assisted Analysis

I use the term "computer-assisted" as a reminder that the computer is fundamentally the same as an abacus, a slide rule or a hand calculator. It is simply another tool that allows for accurate solutions of complex mathematical models of vehicle dynamics. It does not interpret physical evidence, determine what physical principles are applicable or decide what input data is appropriate. It is often, and properly, stated that it does not eliminate the necessity to use the computer between our ears. What it does do is perform mathematical computations really fast.

The preceding chapters have discussed the physics concepts and the mathematical models required for the understanding and description of vehicle and impact dynamics. Analysis procedures for the reconstruction of a variety of impact configurations have also been presented. The use of specific computer programs was not addressed since their use is not actually necessary to understand the reconstruction process. In addition, the quality of computer programs has been improving rapidly in recent years – making many existing programs obsolete.

Although the use of computer programs in accident reconstruction has not been a total success due to their widespread misuse, there are many important advantages to computer-based analyses. These advantages include, but are not limited to the following:

1. Programs can convert direct measurements, surveying output, and photos into accurate and reliable scene and vehicle data, presenting these in an orderly fashion for use by the analyst.

2. Relatively simple programs can be utilized to quickly repeat computations using a range of input variables. For example, changing values of an effective drag factor in a spinout or rollover calculation can establish a speed range that the analyst can use with confidence.

3. More sophisticated approaches can be used to better replicate the actual behavior of the vehicle during various motions such as the post-impact spinout. A consideration of both rotational and translational motion and the inclusion of the effects of steer angle and wheel jamming has the potential for a more accurate result.

4. Speed correlation in the post-impact analysis can be assisted by repeating calculations to allow for a simpler resolution of conflicting speed values for the two vehicles.

5   Analysis of the impact phase can be done by both conservation of momentum and kinetic energy loss due to damage, thus providing an important crosscheck.

6   Kinetic energy losses due to complex damage profiles can be easily computed and stored as a Fixed Barrier Equivalent (FBE) speed. This process can be enhanced by storage of the appropriate vehicle test data.

7   Making small variations in the estimated approach and departure angles can reveal sensitivity problems that might indicate the unreliability of some or all of the reconstruction conclusions.

8   Output from a computer analysis can be presented in an orderly manner, tailored to the users' needs. For example, speed changes in the x and y directions can easily be transposed to lateral and longitudinal ΔV's for use in the analysis of occupant motion.

9   A comprehensive program could perform the entire reconstruction effort from final rest back through any pre-crash maneuvers. Intermediate, as well as final, results could be provided to allow for necessary judgment checks.

10 Mathematical computations are generally more accurate and reliable.

11 The time required to perform the calculation is dramatically reduced.

12 Simulation programs that are not generally efficient for the basic reconstruction process can be quite useful in testing various computer results. This is one good way to find gross errors in input data or analysis assumptions.

13 Animation programs can be quite effective in demonstrating the dynamics of the accident events. This visual display provides another check on the reconstruction result. If it does not *look* right, it probably is not right.

There are, however, some potential problems with using computer programs for accident reconstruction. Programs with a clear and orderly presentation of the output tend to give the user unwarranted confidence in the results. Comprehensive and sophisticated programs are in many ways quite desirable but have several drawbacks.

1   These programs require more input data with the corresponding increased chance for errors.

2   This type of software may not provide intermediate results that can raise warnings regarding input data and assumptions.

3 The more complex a program, the more likely it is that the user does not fully understand the mathematical model and/or the underlying physics. Not understanding the logic of the program can lead to total misuse, resulting in a solution with little relation to reality.

Good practice would indicate that you never use a computer program unless you understand the physics involved, the mathematical models used, the assumptions made and, therefore, the limitations of the analysis.

A final very serious problem arises when people without adequate educational background or practical experience attempt to use a computer program to compensate for their lack of qualifications. In fact the use of computer-assisted reconstruction analysis requires *greater* knowledge of basic principles. The use of hand calculations to give rough estimates before trusting a complex computer analysis is generally a good practice.

Event Data Recorder

One development that could very well revolutionize accident reconstruction and causation analysis is the potential use of a sophisticated "Black Box." Although virtually all manufacturers install some type of event data recorder in all new vehicles, these recorders are not yet satisfactory for reconstruction or causation analyses. The type of data being recorded is generally quite limited and is inconsistent among the various automakers. Retrieval of the data is often costly and cumbersome.

For analysis purposes the recorder should continuously monitor and maintain a cycling storage of events for a short time period leading up to a collision and continuing until final rest. Information that is related directly to accident and injury causation should include:

1 Seat belt usage
2 Air bag deployment
3 Seat and steering wheel positions
4 Headlights, taillights, brake lights and turn signal usage
5 Usage of cellphone, radio, navigation system, etc.
6 Steering input and brake application
7 Tire pressure

Data required for a complete description of the pre-crash and impact phases of the accident include:

1 A digital recording of the pre-crash speed with a sufficient number of data points to properly describe any braking or acceleration maneuvers
2 A plot of steering input versus time
3 A complete recording of the longitudinal and lateral acceleration pulses and the corresponding changes in speed during impact

Post-impact data that would be of definite interest to researchers might include the following:

1 A speed versus time plot to final rest
2 A plot of the center of gravity trajectory
3 The rates of rotation about the x, y and z axes to final rest

Some of this will be discussed later, but how all of the post-impact data would be measured and recorded eludes me. As you probably suspected, I am definitely not an expert in instrumentation.

For accident reconstruction purposes having the vehicle speeds at impact would eliminate much, if not all, of the need for the post-impact motion analysis and of the, often dubious, calculations regarding the impact phase dynamics. It should, however, be recognized that the speed displayed by the event data recorder may not be totally accurate. This speedometer display is predicated upon a calibration for the vehicle traveling in a non-yaw mode with tires of a specific size and tread depth and with a standard inflation pressure. Discrepancies in the speed indicated can occur due to the following:

1 Inherent errors in the system.
2 Significant inflation pressure differences that produce small errors
3 Tread wear will produce recorded speeds that can be higher than the actual speed. For example, a 0.30 inch tread loss on a 12-inch radius tire will produce an error of $0.30/12 = 0.025$ or 2.5 percent
4 Non-standard tires. Using a 15-inch radius tire instead of the original 12-inch tire will create an error $3/12 = 0.25$ or 25 percent. If the recorded speed was 60 mph, the actual speed would have been $60 + (.25) 60 = 75$ mph.

5 A vehicle in a yaw or spinout will have such large errors in speed recording that the readout may be interesting, but is essentially worthless for reconstruction purposes

The errors in 3) and 4) can be accounted for by using the appropriate adjustments.

If the "black box" also has a record of the impact phase of a single-vehicle crash or a two-vehicle collision, this should include the longitudinal and lateral accelerations sustained during impact. The acceleration pulse should include the rate of onset (jerk), the peak value, the shape and the duration. These parameters are dependent upon the impact speed(s), the stiffness of the object or vehicle struck, the angle of impact and, particularly in the case of pole impact, the portion of the vehicle making contact. An approximation of a moderate speed longitudinal acceleration pulse for a striking vehicle is shown in Figure 7.1. The peak rate is approximately 20g's (~ 640 ft/sec^2) and the total duration is about 150 milliseconds (ms).

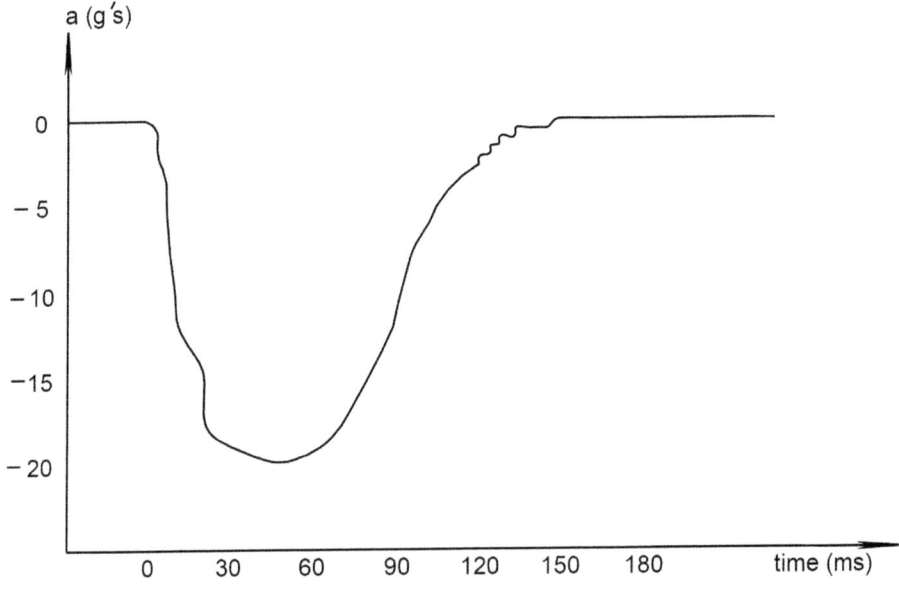

Figure 7.1. Acceleration Pulse

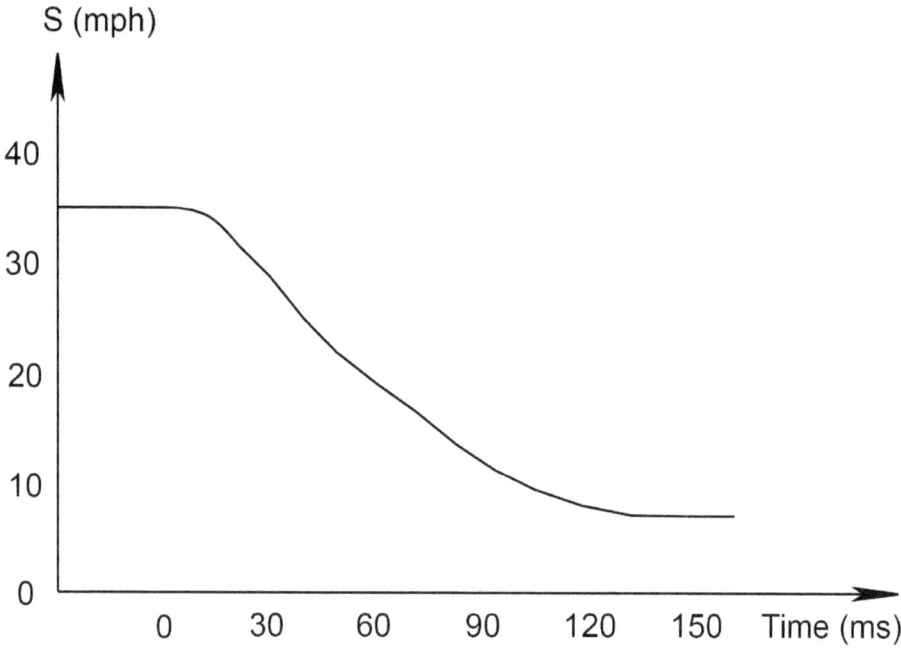

Figure 7.2. Speed vs. Time

Since the change in velocity ΔV or the change in speed ΔS is the integral of the acceleration over the time duration of the impact

$$\Delta V = \int_0^t a.$$

The plot of the speed during the impact phase would be as shown in Figure 7.2. This change in speed could be determined and displayed inside the recorder or by a numerical integration process outside the recorder. Note that this change in speed is equal to the area under the curve as shown in Figure 7.1.

If only one of the accident vehicles has a "black box" and only the impact speed is recorded, this can still be useful. Upon completion of a normal reconstruction this difference between the calculated speed and the recorded (and adjusted) speed can be utilized to make changes in the measured and/or assumed values of effective drag factor, approach and departure angles and impact crush parameters. This will provide an improved, if not completely satisfactory, reconstruction analysis.

If, in addition to the impact speed, this same vehicle had a data recorder that provided the longitudinal and lateral impact acceleration pulse and, therefore, the corresponding changes in speed, further improvements to the reconstruction analysis can be achieved. Both changes in speed of the other vehicle can be determined directly by the use of conservation of momentum. That is, $\Delta S_2 = (W_1 / W_2) \Delta S_1$. This would provide additional criteria that must be met in the adjustment of input data such as effective drag factor, and approach and departure directions.

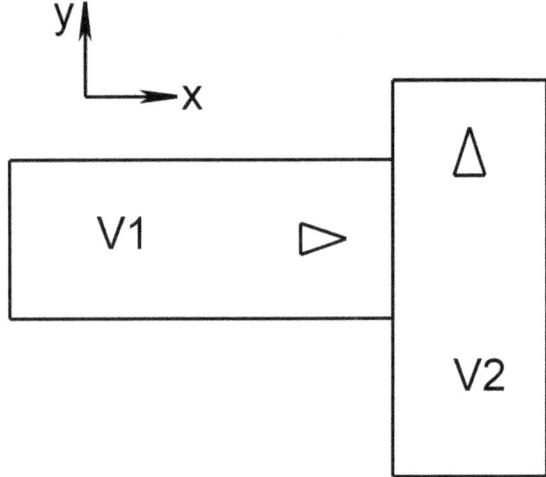

Figure 7.3. Intersection Collision

In some crash configurations a good estimate of the other vehicle's impact speed may be easily obtained. For example, in a typical intersection collision, as shown in Figure 7.3, let us assume that the reorder has provided the impact speed and both changes in speed for Vehicle V2. That is, $S2_1$, $\Delta S2(x)$, and $\Delta S2(y)$ are known values.

The post-impact speed of V2 in the x direction is equal to the change in speed of V2 in the x direction. That is

$$S2_2(x) = \Delta S_2(x).$$

Since this is essentially a plastic impact ($e \approx 0$), the post-impact speeds are approximately equal. That is

$$S1_2(x) \approx S2_2(x) \approx \Delta S2(x).$$

From conservation of momentum

$$\Delta S1(x) = (W_2 / W_1) \Delta S2(x)$$

and

$$S1_1 = S1_2(x) + \Delta S1(x)$$

or

$$S1_1 = \Delta S2(x) + (W_2 / W_1) \Delta S2(x)$$

or

$$S1_1 = (1 + W_2 / W_1) \Delta S2(x).$$

Note that in this perpendicular collision the change of speed of V2 in the y direction, $\Delta S2(y)$ is not needed.

If the impact speed and changes in speed of V1 only are known in this same crash configuration, the changes in speed of V2 can be found by conservation of momentum, but the impact speed of V2 cannot be found as easily. Known values now would include $S1_1$, $\Delta S1(x)$, $\Delta S1(y)$, $\Delta S2(x)$ and $\Delta S2(y)$. The impact speed of V2 in the x direction would be equal to zero, but the y component of V2's impact speed $S2_1(y)$ is still unknown.

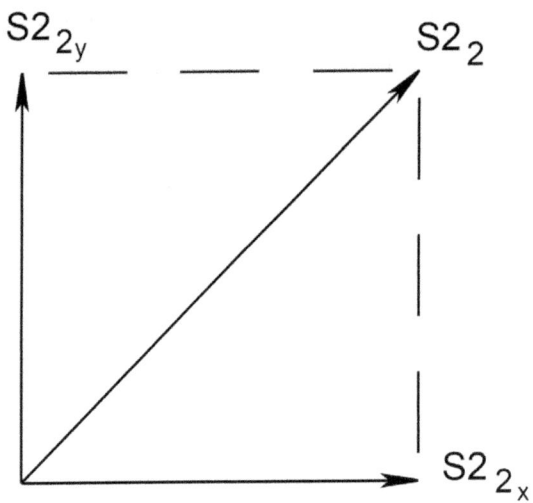

Figure 7.4. Post Impact Speed

A simple way to determine what other information is required is to use a graphical display of the post-impact speed $S2_2$ and its components as shown in Figure 7.4. Since $S2_2(x)$ is known, the rectangle and, therefore, the magnitude of $S2_2(y)$ can be completed if we can, from physical evidence, determine the magnitude of $S2_2$ or the departure direction of V2. If we have estimates of both, this will give a range of values for $S2_2(y)$ and, therefore, $S2_1$ which is given by $S2_1 = S2_2(y) + \Delta S2(y)$.

In all collision configurations, if at least one of the vehicles has the recording of the acceleration pulses and changes in speed, consideration of damage is now unnecessary since all impact changes in speed can be found by conservation of momentum. For example, a head-on collision could be reconstructed if one of the post-impact longitudinal speeds can be determined along with the impact speed of one of the vehicles. Another example would be a rear-end collision with a stopped vehicle. The stopped vehicle's change in speed will directly yield the post-impact speed of both vehicles and the impact speed of the striking vehicle.

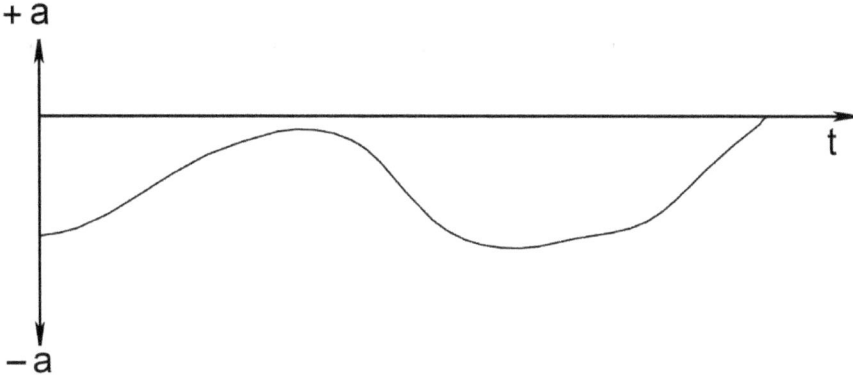

Figure 7.5. Spinout Acceleration Pulse

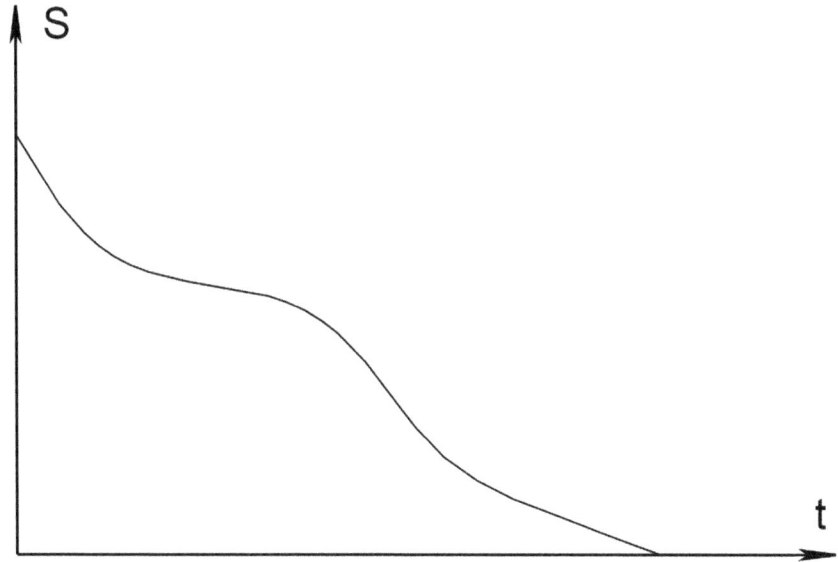

Figure 7.6. Spinout Speed Loss

The determination of the speed loss in a yaw or spinout could possibly be determined by the use of sensitive accelerometers that record the lateral and longitudinal decelerations, thus yielding the speed change versus time to a subsequent impact, a rollover or final rest. Examples of the possible plots are shown in Figures 7.5 and 7.6. These plots can then be correlated with the physical scene evidence. This same procedure might be applied to rollovers but, as I have said before, I have no useful suggestions as to exactly how the instrumentation and measurement could be carried out. This information would be particularly useful in the analysis of post-impact motions and single-vehicle accidents.

In two-vehicle crashes, if both impact speeds and all speed changes are recorded, conservation of momentum should always be used to provide a cross check for possible errors in either data recorder. This can be done by converting all changes in speed to a common $x-y$ coordinate system and then determining if the following relationships are satisfied

$$\Delta S1(x) = (W_2/W_1)\Delta S2(x)$$

$$\Delta S1(y) = (W_2/W_1)\Delta S2(y).$$

In all types of accidents the displayed data should also be correlated with the physical evidence. If possible, good practice would be to utilize the available vehicle and scene evidence to conduct a traditional reconstruction analysis for comparison purposes.

Having the output of a pre-crash speed versus time will reveal not only the pre-crash initial speed but also the magnitude and timing of any braking or acceleration maneuvers. A similar plot of steering input and/or lateral acceleration will define any steering avoidance maneuver. A timeline showing the location of each pre-crash event can be displayed on a drawing or aerial photo, or marked at the actual accident site.

This information will allow for an evaluation of the driver's actions that might have contributed to the cause of the accident such as speeding, inattention or a slow or inappropriate response. Correlation with the accident scene layout and physical evidence can also help identify any driver/highway interaction causal factors.

In addition to significant improvements in the reconstruction and accident causation analyses, the event data recorder will also benefit the evaluation of injury causation mechanisms. This information can also be extremely useful in research efforts to better understand vehicle dynamics in pre-crash and post-crash phases of an accident. Impact acceleration pulses are already used in the understanding of the effect of crash severity (ΔV, etc.)

upon vehicle crush and injury severity. This improved understanding makes the development of better vehicle control and stability possible, as well as improved crash survivability characteristics.

Privacy concerns will undoubtedly be an issue in the use of "black box" data. In research activities privacy can be maintained in the same fashion that medical data is sanitized by eliminating names, dates, locations and other unneeded information. Using these data in criminal cases and civil litigation will produce a better justice system, allow for a more fair resolution of disputes, and assist in counter-measures against culpable drivers and other responsible parties.

A strong argument can be made that, if I, as a vehicle owner or driver, am going to use a publicly owned, financed, and operated highway system, the public and the courts are entitled to this information. Legislatures and the courts will ultimately decide on the limitations on the type of data collected and how it is used.

Finally, it is important to recognize that the availability of "black box" data does not eliminate or reduce the need for qualified, experienced professionals to interpret and properly utilize this information. The interrelationships between drivers, vehicles and the environment will still have to be determined in order to complete the reconstruction and causal analyses. Injury causation mechanisms still have to be evaluated by biomechanical engineers and medical experts.

Hopefully, more emphasis can be directed toward determination of accident and injury causation factors and the development and implementation of effective countermeasures that can actually reduce the frequency of highway crashes and the severity of injuries.

Research will undoubtedly continue to contribute to our understanding of all aspects of vehicle dynamics, impact dynamics and vehicle stability and control. Additional testing with improved instrumentation can lead to improved mathematical models for spinouts and rollovers and a better grasp of changes in speed and acceleration during impacts, as well as the collection of data that we may not even have imagined. More comprehensive crash testing will provide better crush coefficients for a variety of impact configurations. This, along with ever-improving computer programs, will allow for more reliable reconstruction methodology.

Causation Summary 8

Although accident causation has been briefly mentioned in previous chapters, a more thorough discussion is now appropriate. This chapter provides a methodology for the identification of accident causation, as well as a description of specific causal factors. A non-medical view of injury causation is included at the end of this chapter.

Qualitative and quantitative results from the reconstruction effort, although often incomplete, form the basis for determining the causal factor(s) leading to the accident event. This causal analysis is, of course, the ultimate objective of the accident investigation and reconstruction process. The identification of causal factors is facilitated by considering the operation of a motor vehicle as a continuing series of accident avoidance maneuvers. This process that might or might not be successful consists of several phases as shown in Figure 8.1.

206 | HIGHWAY ACCIDENTS

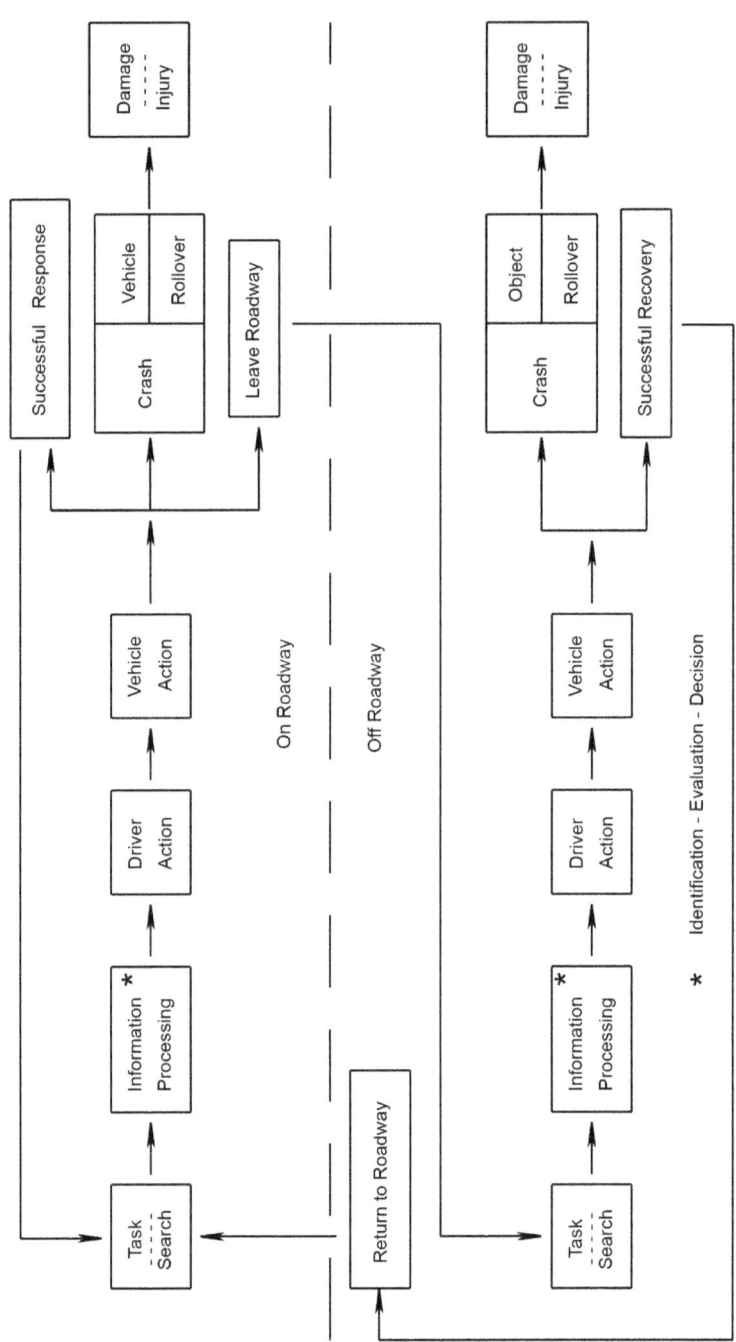

Figure 8.1. Accident Causation Model

The task consists of the driver using the appropriate speed, maintaining the proper course and carrying out the required Search as described in Chapter 2 Evidence, Reaction Time. She will be continuously faced with a situation such as a curve, a stop sign or some other condition requiring a response. Each new situation prompts the Information Processing step that consists of the Identification, Evaluation and Decision phases of the reaction process.

Driver action is then required to execute the Decision from the Information Processing. This is followed by a Vehicle Action resulting in a successful response, a crash occurrence, or a roadway departure. If the vehicle leaves the roadway, a new off-road situation will initiate the same process leading to a successful return to the roadway or a crash.

Failure in one or more of the phases in this process can generally be identified as causing the accident. Causal factors may be classified as human, vehicular, environmental or some combination.

Human Factors

The majority of accidents are caused by driver failure(s) in the Search through Driver Action process. There may be predisposing conditions that compromise proper driving behavior throughout the entire process. These would include permanent conditions such as physical and mental deficiencies, as well as the more frequent temporary conditions such as impairment, illness and fatigue.

Some of the common human causal factors listed in a rough order of importance are as follows:

1 **Inattention:** This failure in the Search mode is probably the most prominent cause of roadway collisions.
2 **Distraction:** Another common and similar failure in the Search is being distracted by non-driving activities. These include looking at outside events, eating, drinking, interacting with passengers, and especially talking and texting on the ubiquitous cell phone.
3 **Impairment:** The use of drugs, especially alcohol, can have a negative effect upon all phases of driving and is a major causal factor in serious and fatal crashes.

4. **Speeding:** Excessive speed is not only the cause of many collisions but also greatly increases the crash severity.
5. **Reckless Maneuvers:** Failure to yield, tailgating, sudden speed changes, weaving through traffic, and other dangerous maneuvers actually create accident-causing situations.
6. **Emotional Instability:** Road rage is often a factor in leading a driver to make really bad decisions.
7. **Inexperience:** New vehicle operators, as expected, have difficulty in evaluating situations and making good decisions. Their lack of basic driving skills, particularly for new motorcyclists, leads to driver actions that are too slow, inappropriate or excessive.

Speeding, failure to yield, and tailgating are examples of driver actions that can be the initiating factor leading to an accident event. Excessive speed decreases the driver's ability to successfully execute steering and braking avoidance maneuvers and to safely negotiate curves. Speed in excess of the posted limit can lead to another driver misjudging his ability to execute a crossing or passing or lane change maneuver.

Failure to yield by not stopping for a right turn on red, overrunning an amber warning signal, or failure to stop for a stop sign or red traffic signal are obvious direct causes of collisions. Avoiding a vehicle that ran a red light or stop sign is somewhat more complicated.

Most drivers approaching a stop sign or red light will brake using an effective friction factor of 0.2 to 0.4 (braking at 0.2 + 0.4 g's). Since this normal driving behavior is part of our everyday experience, vehicles not braking until they are close to the intersection fall within our expectations and cause no concern. If, however, a driver fails to brake and continues into the intersection, should you be expected to avoid a collision? To analyze this let's use a hypothetical situation where you are driving V1 at the speed limit and V2 is approaching from the right, also at the speed limit. The pertinent conditions are as follows:

- The speed limit is 30 mph (44 feet/second) in both directions
- The pavement is dry with $f = 0.8$
- The sight distance is unlimited
- Each vehicle is 6 feet wide and 15 feet long

As V2-1 gets to the location where he must brake at $f_e = 0.4$ (0.4 g's) to avoid impact, he is a distance

$$d_2 = (30)^2 / 30(.4) = 75 \text{ feet}$$

from the potential POC. At this time you (V1-1) are into the evaluation phase of your reaction process. By the time V2 passes the point where the braking required is $f_e = 0.6$ (hard braking), he is now only a distance equal to

$$d_2 = (30)^2 / 30(.6) = 50 \text{ feet}.$$

from the conflict point and you should be deciding to brake to avoid him. You have been tracking him for a distance of 25 feet and a time equal to

$$t = 25 \text{ ft}/44 \text{ ft/sec} = 0.57 \text{ sec}$$

or approximately one-half second.

Since you have been alerted and are through the evaluation phase, your remaining reaction time should be no more than about $t_R = 1.0$ second, rather than 1.5 to 2.0 seconds. During your 1.0 second reaction time V2 will have moved 44 feet, and is now a distance

$$d = 50 - 44 = 6 \text{ ft}$$

from the potential impact.

Your ability to avoid V2 is now solely dependent upon where you happen to be relative to its location. Your braking distance is approximately equal to

$$d_B = (30)^2 / 30(0.8) = 38 \text{ ft}$$

and the braking time is equal to

$$t_B = V / a = 44 / 32(.8) = 1.7 \text{ sec}.$$

The total stopping distance d_T would be

$$d_T = 44 + 38 = 82 \text{ ft}$$

and the total time to react and brake would be

$$t_T = t_R + t_B = 1.0 + 1.7 = 2.7 \text{ sec}.$$

If you are a distance $d > 82$ feet from the point of conflict, you could react and brake to a stop prior to the POC.

During this time interval of 2.7 sec, V2 will have traveled a distance of

$$d = 44(2.7) = 119 \text{ ft}.$$

The distance required for V2 to clear your travel path is equal to

$$d = 50 \text{ ft (to POC)} + 6 \text{ ft (your car width)} + 16 \text{ ft (his car length)} = 72 \text{ ft}.$$

Therefore, V2 would have cleared your path by 47 ft. or slightly more than one second. It is clear there is some distance less than 82 ft. (not calculated here) that you could react, brake and slow sufficiently to allow V2 to clear your path.

In general the speed limit on the through street will be greater than on the stop street and it will be even more difficult to avoid a collision. This higher speed limit also makes it more difficult to avoid a vehicle coming from a driveway at a slow speed. If V2 is approaching at 10 mph (~15 feet/second), braking at $f = 0.4$ requires

$$d_2 = (10)^2 / 30(.4) = 8 \text{ ft}$$

and at $f = 0.6$

$$d_2 = (10)^2 / 30(.6) = 6 \text{ ft}.$$

Since these decision points are crossed at approximately one-half second or less, if you are close to the intersection, you do not have time to react, let alone brake.

In summary, if you are in the wrong place at the wrong time, by the time you realize there is a problem, it is generally too late to avoid a collision.

Rear-end collisions, a common type of accident, occur most frequently in urban areas, particularly during traffic rush hours. Following too closely (tailgating) and inattention are the two major causes. In order to address the problem, states have issued Driver Handbooks containing "rules of thumb" for safe following distances. Most commonly expressed are the following:

1 Keep your following distance to at least one car length per 10 mph.
2 Maintain a two-second gap behind the vehicle in front of you.

As we shall see shortly, these two rules sometimes found in the same handbook are contradictory.

The basic avoidance requirement is this: when the vehicle in front of you brakes hard, you should be able to react and start braking before the front of your vehicle reaches the point on the roadway where the rear of the leading vehicle was when it began braking. This can be achieved only if you are following at a time interval greater than your reaction time and, of course, are paying attention to your driving.

Using the first rule, consisting of a gap of one car length (~ 15 feet) per 10 mph (~15 feet/second), translates to a following interval of approximately one second. That is, your required reaction time is ≤ 1.0 second. Since the range of reaction time for most people most of the time varies from 0.5 second to 1.5 second, with an average of approximately 1.0 second, we must react *better* than our average reaction time.

Since using rule #1 will inevitably lead to hitting someone, it should, in my opinion, be discarded as inadequate. Rule #2, a two-second gap, is much more reasonable but still does not provide a great margin of safety.

Some people argue that if they maintain a two-second gap in heavy traffic on a multilane highway, somebody will change lanes in front of them and eliminate the safe gap. Since this lane changer is probably going to maintain a gap of ≤ 1.0 second, they would be required to drop back only 1-2 seconds.

If this should happen to you 10 times on your commute to work, you would arrive 10-20 seconds later, but probably a whole lot less stressed.

Some aggressive (this sounds better than rude and reckless) drivers will habitually move up to one half a car length or less in heavy 60 mph traffic. This translates to a gap of < 0.10 seconds. Therefore, if the vehicle in front brakes, there will be zero possibility of avoiding a collision, regardless of the driver's often-inflated sense of his reaction capability.

It is important that motorcyclists maintain a greater distance since, contrary to some riders' opinions, motorcycles often do not stop as quickly as automobiles. Motorcyclists should also keep in mind the consequences if they do rear-end another vehicle. Occasionally we see a tractor-trailer tailgating a car at highway speed. This is particularly irresponsible since a truck's stopping capability is much less than a car. Being rear-ended by a tractor-trailer has a disastrous potential.

Although most people are really quite poor at judging the actual time and distance, the analog computer between our ears, along with experience, can assist us in judging our following distance. Keeping a safe following distance does not, however, guarantee that you will not rear-end someone. You still must maintain attention on the roadway ahead of you. I have had some close calls because my mind was focused on other things (such as what I should tell other people about safe driving) and failed to pay proper attention to my driving task.

Failures in the Search and Identification phases are generally fairly easy to define, but Evaluation errors are often more subtle. Proper estimates of times, distances and speeds, although never done numerically, are necessary to evaluate situations that might require a driver's response. This mental computing skill can only be learned by actual driving experience. Conditions that are unusual or confusing can also cause experienced drivers to improperly assess the situation.

Decision errors often involve the choice between steering and braking. At times a simple steering input is all that is required and panic braking is not useful. In situations where an impact is inevitable, attempting to steer through the problem is the wrong choice since braking will at least reduce the consequences of the crash. A decision to steer and brake requires carefully applied brake pressure if you don't have functioning ABS brakes. In a panic situation it is easy to lock up the front tires and loose all steering capability.

Older drivers need to accustom themselves to the ABS braking system so they will not hesitate to brake and steer.

Although experienced drivers generally make better, as well as quicker, decisions, there are exceptions. An example of this was a local driver, traveling westbound on a narrow, unlined rural road, coming over the crest of a hill and encountering an eastbound vehicle headed directly toward her. She responded properly by moving quickly to her right but the other driver swerved to his left and struck her head-on. The ensuing investigation into the cause of this head-on collision that clearly occurred in the westbound lane initially focused on the suspicion that the eastbound driver was fatigued or impaired. He was in fact a sober and alert businessman who had just arrived from England and responded by instinctively moving to the left based upon *his* prior driving experience.

Failures in the Driver Action phase are often due to a slow response but can be due to a loss of steering through excessive braking, as previously described. A frequent cause of loss of control is a steering over-correction following a lane departure due to an emergency avoidance steering maneuver. This overcorrection is not necessarily a sign of bad driving, but simply a driver who is not experienced with a particular situation

Three groups of drivers receive special concern and criticism – the old, the young and the repeat offender. Some elderly drivers demonstrate poor driving behavior due to physical impairments and mental limitations such as lack of concentration. Although they should probably not be driving and are hard to identify, they do not represent a significant percentage of the overall travel mileage.

Young people are significantly overrepresented in vehicular crashes. While lack of driving experience is often the reason for their involvement in an accident, many are simply prone to irresponsible behavior. According to behavioral scientists one explanation might be a developmentally explained inability to adequately perceive the consequences of their actions and, thus, their feelings of invulnerability. Fortunately, most will outgrow these tendencies.

Repeat offenders are a significant problem. Evidence shows that a small percentage of drivers are involved in a large percentage of severe crashes. Some of their common driving behaviors include:

- Driving while impaired
- Speeding

- Hard braking and acceleration
- Failing to use turn signals
- "Cutting off" other vehicles
- Frequent lane changing
- Tailgating
- Ignoring traffic control devices

The identification and understanding of human causal factors often requires an evaluation by a human factors expert.

Vehicular Factors

Accidents caused by vehicular problems comprise a small percentage of the total but can be instrumental in causing serious and fatal crashes. Common vehicle problem areas that contribute to accidents include the following:

- **Lights.** Inoperative headlights, tail lights, brake lights and turn signals that are essential for safe vehicle operation.
- **Windshield Wipers.** Worn wiper blades interfere with driver visibility and lead to failures in the Search phase.
- **Brakes.** Worn or misadjusted brakes can basically eliminate a driver's ability to perform an avoidance maneuver. This is an all too common problem with trucks operated by independent driver/owners who are paid by the load.
- **Tires.** Worn tires with little or no tread seriously degrade the ability to steer and brake in wet weather and are a major cause of hydroplaning. Blow outs and tread separations often lead to loss of control since most drivers do not know how to respond and have never practiced coping with this problem.

A thorough vehicle inspection is essential to identify these causal factors. Other vehicle problems due to obsolescence or the result of bad design decisions include:

- A short wheel base (tending to cause over-steering)
- A high rollover tendency

- Dynamic instability
- Limited driver visibility

Identification of rollover and stability issues generally requires input from a vehicle specialist.

Environmental Factors

Undesirable situations caused by rain, sleet, snow, ice, wind, fog and smoke are obviously uncontrollable but become a serious problem when drivers fail to adjust their driving behavior to these ambient situations. Vehicle maintenance deficiencies and inadequate highway design, construction, maintenance or operation practices can also make adverse weather conditions even more hazardous. Highway design, construction, maintenance and operation practices that contribute to accidents are numerous, serious and often complex. A brief list of some simple and obvious defects include:

- Potholes and other pavement discontinuities
- Pavement/shoulder drop-offs
- Blind entrances
- Faded and obscured signs
- Malfunctioning traffic signals

Other more subtle defects include:

- Worn slick pavement
- Inadequate drainage of travel lanes
- Narrow or non-existent shoulders and roadsides
- Shoulders that have not been stabilized
- Unguarded roadside hazards
- Sudden termination of a lane
- Inadequate distance for required merges and other lane change maneuvers
- Sharp horizontal curves without a proper spiral to accomplish the superelevation transition

- Inadequate sight distance at intersections or on approaches to stop signs, traffic signals, and changes in vertical or horizontal alignment
- Improper or misplaced signage

Highway engineering expertise is required to identify these deficiencies. Other highway defects that interact with the driver will be discussed later.

Causal Interactions

Identification of the "at fault" driver is the usual goal of the investigating police officer. This is understandable since issuing citations and making felony arrests are an essential part of their job. This approach, however, suggests that only one driver is responsible for an accident and is certainly not adequate for identifying other causal factors. "Driver error" should not be the end of the analysis but the beginning of a search for more subtle interactive causal mechanisms. An accident analyst's task is to find all of the human, vehicular and environmental factors that contributed to the crash.

A common example of multiple driver involvement occurs when a car enters the roadway from a stop sign and is struck by an oncoming vehicle. The driver is cited for "failure to yield" and that is usually the end of the investigation. Further investigation may reveal that the other driver would have been able to avoid the collision had he not been speeding. Secondary human causal factors are often ignored, resulting in an incomplete causal analysis. An example I recall was a near-sighted driver who was not wearing his glasses, driving at night in the rain, trying to see through a windshield with worn wiper blades. Not surprisingly, he missed a stop sign and collided with another vehicle. This accident occurred due to a combination of human and vehicular causes.

Some accidents are the result of several interrelated factors. I remember a case where a woman was driving a car with bald tires in the rain at a high rate of speed on a street with worn pavement and inadequate drainage. She hydroplaned, lost control and struck a jaywalking pedestrian. Here we have two human, one vehicular and two highway factors (three if you count the rain) working together to cause the accident.

A less obvious case of driver/highway interaction involved a driver traveling through a multilane complex interchange. He passed a series of closely spaced signs giving directions and, while trying to process this information, he missed an exit from the *left* lane that was inconsistent with his normal expectations. He then tried to quickly change lanes, over-steered and struck a concrete barrier.

There are many similar examples of driver-highway interaction that require the analyst to consider that all drivers have limitations in the information processing stage. These include:

- Complex situations can be difficult to understand and, therefore, increase a driver's reaction time in the evaluation and decision stages of the reaction process.
- Drivers are limited in how fast they can receive and absorb information and directions.
- Rapid decision-making requirements lead to driver errors.
- Inconsistent design features and signage are a major cause of "driver errors" – driver expectations are important

Interaction between the driver and the environment created by rain, fog or smoke obviously can negatively affect the identification phase of information processing. This phase can also be affected by problems with highway lighting, headlights and background lights. In order to properly identify and then evaluate a potential problem the driver must, of course, be able to clearly observe it.

Inconsistencies in highway lighting that can create difficulties include the following:

1. Entry to an unlit tunnel or garage from bright daylight.
2. Sudden (rather than gradual) termination of high level roadway lighting.
3. Street lights spaced too far apart, thereby creating dark spots.
4. Street lights blocked by vegetation, thus creating dark spots.

Since it takes time for eyes to adjust, the sudden loss in lighting intensity will leave the vehicle operator driving virtually blind.

Identifying tail lights, brake lights or headlights can be inhibited by bright background lighting from adjacent property or from highway construction activity. On-coming headlights tend to produce a relative blind spot adjacent to them, thus obscuring a driver's view of any potential hazards in his lane of travel. Traveling at speeds in excess of 30 to 40 mph while using low beams places the driver in the position of "over driving the headlights." That is, by the time the driver observes a pedestrian, an object or a hazard, it is too late to react and execute a successful avoidance maneuver. Driving directly into a sunrise or sunset can effectively blind a driver, thus eliminating any successful identification process.

A typical problem in the evaluation phase is the difficulty drivers encounter when trying to determine the speed of vehicles traveling directly toward them. This is much worse at night since drivers lose sight of surrounding objects for comparison purposes. Judging the speed of approaching headlights is a difficult task for even the most experienced drivers.

It is also difficult at night to determine if an approaching vehicle is in your lane, particularly on a curve. It is even harder to judge the speed or location of a motorcycle based upon an observation of its single headlight.

If one of more of these above-described environmental situations is present, it must be taken into consideration in our evaluation of "driver error."

Road construction, maintenance and other operations cause a disruption of the free flow of traffic and produce situations that are annoying and inherently dangerous. Creation of confusion and unnecessary hazards by construction contractors, highway agencies and utility companies is widespread and is a significant factor in the causation of many serious and fatal crashes.

A great deal of research and effort has gone into the development of safety standards for construction activity. There is a *Manual of Uniform Traffic Control Devices*, a federal standard adopted by all 50 states, as well as many state and local standards. Failure to follow these *required* standards along with deviating from accepted engineering practice is seen all too frequently. One motivation appears to be economic, since omissions and shortcuts can increase profits or reduce losses when a bid was too low. A common shortcut is when contractors do not have someone on their staff with adequate knowledge regarding standards and safe practices, let alone accident causation.

Highway agencies spend millions of dollars to design and construct upgrades to our highway system but often fail to properly supervise the construction process. This supervision may be directed mainly toward the assurance of the quality of the final product with insufficient concern for the safety of motorists and pedestrians traveling through the construction site. Inspectors often do not have the time to identify and correct unsafe conditions. They rarely have the qualifications to recognize situations that are potential driver-highway accident causal factors. Common examples of direct hazards include:

- Construction equipment parked unnecessarily close to the roadway
- Excavations
- Severe pavement/shoulder drop-offs
- Lack of temporary barriers.

Frequent deficiencies that contribute to the creation of potential driver-highway accidents are the following:

- Lack of advance warning signs
- Lack of signs indicating what the motorist is facing and what action is required
- Misplaced or inappropriate signs (see Figure 8.2)
- Inadequate guidance (pavement markings, barricades, cones, etc.) into and through construction zones
- Failure to remove old markings and signs that are in conflict with temporary routes (see Figure 8.3)
- Inadequate or misplaced temporary traffic signals
- No flagmen or police assistance for transient activities
- Failure to provide a safe pathway for pedestrians

The causal factors attributing to the interaction between the vehicle and the highway have generally been addressed in our discussion of vehicle dynamics and accident reconstruction.

Figure 8.2. Misplaced Sign (On Ground)

Figure 8.3. Conflicting Markings

Injury Causation

Although the determination of injury causation mechanisms requires expertise from the medical and biomechanics fields, vehicle specialists and reconstruction analysts can also provide valuable contributions. Human, vehicular and environmental factors contribute to injury causation.

Failure by the driver or the passengers to properly utilize the available restraint system is the primary human factor that increases the severity of injuries. Unrestrained occupants can be ejected, as well as impacting the interior of the vehicle. Reconstruction of the occupant kinematics, as described in Chapter 4 Impact Dynamics, Occupant Dynamics, is the first step in most injury causation analyses.

Vehicular injury causation factors include, but are not limited to, the following:

- Inoperative or inadequate seat belts, air bags, head rests, or child restraints
- Inadequate interior space to allow the required occupant movement without striking the vehicle interior
- Lack of occupant compartment integrity, such as roof collapse or a side door intrusion
- Doors, sunroofs, windshields, backlights or side windows that allow the ejection of unrestrained occupants. Since injury severity is, in most cases, significantly greater for ejected occupants, being "thrown clear of the wreck" is not a desirable outcome
- Fire

Vehicle fires are occurring less frequently, but remain a problem that can produce horrific injuries and fatalities, particularly for tankers and trucks with external fuel tanks. Improvements in design have eliminated a number of past design defects including:

- Rear bumper supports that penetrate the gas tank in a rear-end collision.
- Placing the car battery next to the fuel tank in the front of a rear engine car.
- Filler pipes that can easily break loose and spew fuel into the passenger compartment.
- Use of flammable materials in the vehicle occupant compartment.

Although a fire may start with an audible "whoosh," the violent explosion following the start of a vehicle fire as shown in movie scenes requires the assistance of something like dynamite. Hydrogen-fueled vehicles will also not explode, contrary to the hype from certain special interests.

Tracing the origin and causes of fires is a specialized field of science with only a few experts. I am not one of them.

Countermeasures

The identification of accident and injury causation mechanisms is of no use if steps aren't taken to alleviate the problems. In most cases it takes a governmental response in the form of judicial or legislative action to control driving behavior, improve vehicle design and maintenance, provide financial resources for upgrading the highway network, and properly supervise construction and other roadway activity. In some cases criminal prosecutions and civil lawsuits to penalize those responsible for contributing to the causation accidents and/or injuries is appropriate. Injury mitigation will be discussed later in this chapter. Countermeasures for reducing the frequency of accidents will be addressed by first looking at driver regulation.

Drivers

Accident investigation research carried out over the years shows that the most frequent cause of failures in the operation of the highway system is inappropriate driving behavior. For the purpose of suggesting countermeasures, it is useful to classify drivers as roughly belonging to four different (but overlapping) groups – the elderly, the young, the problem, and the normal. Elderly drivers are generally experienced and responsible people who upon reaching a certain age, that is clearly different for everyone, develop physical and/or mental deficiencies that render them unable to drive safely. I am not suggesting any arbitrary age limit discrimination but some of us would be well advised to eventually give up driving. Retesting appears to be a fair and effective method for identifying unfit drivers of all ages, particularly since it has been shown that the mere existence of a retesting system will cause many people to voluntarily give up their licenses.

Young drivers generally are not as skillful and obviously do not have the experience needed for competent driving. It is interesting to note, however, that the high school driver training programs, contrary to what seems logical, have not been effective in reducing the crash frequency and severity among young drivers. Although it is unknown if a more intensive and rigorous program of driver training would be effective, it appears that social and personal responsibility may be more important than knowledge and skill. Another problem with young drivers may is their apparent failure to fully comprehend the potential consequences of their actions. This feeling of being invulnerable or "bullet proof" makes for a brave and effective soldier, but is definitely not a desirable characteristic for operating a motor vehicle. A probation period for young drivers and new drivers is a countermeasure with some apparent success when it is rigorously enforced. An automatic suspension (e.g., from age 16 to 18) for willful driving offenses such as speeding, tailgating or other reckless maneuvering shows some promise of being an effective deterrent to unacceptable driving. One positive solution is that most young people will, if they survive, outgrow this irresponsible behavior.

Problem drivers represent about ten percent of all drivers, but are responsible for approximately one-half of all serious and fatal crashes. Although not covered by the Second Amendment, owning and operating a 3000-pound deadly weapon is considered by many Americans to be an absolute right. One fact to keep in mind is that in most states in this country vehicles still kill more people than firearms. There have been efforts to revoke the license of any driver displaying a pattern of speeding, reckless driving, driving while impaired or other willful acts of irresponsible driving behavior. Even though typically considered a felony offense, some of these people will continue to drive without a license regardless of the consequences. Their long record of driving infractions and "accidents" is consistent with other antisocial behaviors such as problems holding a job, paying their debts, domestic violence, illegal activities and drug abuse. Motorcyclists exhibiting this kind of behavior (e.g., high speed maneuvers such as "wheelies") may callously be considered a self-correcting problem.

The persistent problem of anti-social driver behavior is a real challenge for our society. Our record of lack of success in curbing similar problems such as theft, assault, family abuse and gun violence does not provide great optimism for success. It is hoped that psychologists, sociologists, police officers, judges and legislators can develop some feasible and effective countermeasures. If meaningful action is not taken, these problem drivers can be expected to continue to kill hundreds of people every month.

A "normal" driver can be defined as a reasonably responsible individual who generally drives at moderate speeds, complies with traffic control devices and is considerate of other drivers and pedestrians. Current licensing criteria involving knowledge of basic traffic laws, "rules of the road", and basic operating skills eliminates very few prospective drivers. An example of this was a young lady explaining why an intersection collision was not her fault. "I slowed down for the flashing red light. The other driver ran the yellow and hit me."

Given the very low threshold criteria for obtaining a driver's license it's surprising that there aren't more accidents. Adherence to traffic laws is primarily due to voluntary compliance since police enforcement is minimal with most traffic infractions going unnoticed. Concentration of enforcement efforts should be, and generally is, directed toward the following infractions:

- Speeding
- Reckless driving
- Diving while impaired
- Severe tailgating
- Disregarding traffic control devices
- Texting and other unnecessary distractions

Normal drivers will generally respond to enforcement activities that can result in fines, suspensions and possible jail time by stricter adherence to traffic laws and safe driving practices.

Vehicles

Evolving improvements in the design and manufacture of all types of vehicles have significantly reduced the role of the vehicle in contributing to accident causation. These include:

- Improved tires due to better materials, design procedures, and manufacturing techniques have yielded better tread designs and a prolonged useful lifespan. The frequency of blowouts and detreading has been, with few notable exceptions, significantly reduced.
- Hydraulic and pneumatic antilock braking systems have not only improved braking efficiency (higher effective drag factors), but also allow for a safe combination of steering and braking. Motorcyclists can apply full braking and steering without losing their upright stability.
- The virtual elimination of steering failures and slack in the system, along with power steering, allow for easy and precise steering.
- Vehicle control and stability has been greatly enhanced by improved suspension systems, all wheel drive (AWD), and automatic power distribution.
- Improved driver visibility
- Sensor-assisted driving

Evolution of the latter two categories of improvement will be discussed in some detail later.

In order to fully benefit from these vehicle improvements, it is necessary to regularly inspect and properly maintain these safety features. Mandatory inspection programs have had some positive effects, but have been resisted by the driving public and state officials. It is reasonable to expect that improvements in vehicles will continue in the future through both regulatory mandates and voluntary innovations by vehicle manufacturers.

Driver visibility, as limited by the structure of both trucks and passenger vehicles, has been a continuing problem adversely affecting such maneuvers as backing, changing lanes and parking. Over the past century there have been significant improvements introduced to reduce collisions with other vehicles, pedestrians and objects that are not readily noticed by the driver.

COUNTERMEASURES | 227

An early attempt to improve safety for the backing maneuver was the installation of interior rearview mirrors in passenger cars, but these were useless for many trucks. Outside mirrors were a significant addition, but still left a blind spot immediately behind large trucks. Back-up cameras can alleviate this problem both for automobiles and trucks.

There is also a blind spot in front of cars, buses and trucks that can vary from a height of less than 3 feet to a height that blocks an adult pedestrian from view. This problem that has contributed to many fatal pedestrian accidents and can be alleviated by appropriately placed mirrors.

A parallel vehicle that has a slight overlap with the rear of a passenger car can easily go unnoticed. In the investigation of a lane change accident, matching the two damage locations might reveal that this visibility limitation, in addition to driver error, contributed to the cause of the collision. A good driving tip is to stay out of potential blind spots by either backing off or continuing past the other vehicle.

Figure 9.1. Mirrors – Exterior View

Figure 9.2. Mirrors – Interior View

A classic accident illustrating visibility problems is when a tractor-trailer driver turning to the right in an attempted lane change maneuver, strikes a passenger car located to the right and slightly forward of the cab. Over the years additional safety features, as shown in Figures 9.1 and 9.2, have reduced the frequency of this type of collision. The following changes have been introduced in this general sequence:

1. Lengthening of the right side mirror
2. Placement of a circular convex wide-angle mirror below the flat mirror
3. Locating a second wide-angle mirror on the right front fender
4. Inserting a small window in the bottom of the right door

Although the incorporation of these mirrors, windows, and cameras improve driver capabilities for various maneuvers, they must still be properly monitored by the driver. Driver visibility remains a factor to be considered in the investigation and analysis of many collisions.

Sensor-assisted driving is developing at a rapid pace with many useful features available now or in the near future. These features, include, but are not limited to, the following:

- Alerting a driver that a vehicle is in her "blind spot"
- Alerting a driver that he is speeding, drifting out of his lane, tailgating, or dozing off
- Providing an internal and external alarm when being tailgated
- Warning of upcoming traffic signals, stop signs, speed limit changes, and lane change requirements

Research and development is underway that will use sensors and computers to actually take over some, or even all, of the driving functions. Some vehicles are now capable of such tasks as automatically slowing when following too closely and parking without driver assistance.

If the goal of a completely self-driving vehicle could be achieved, some potential benefits would include:

- Increased traffic efficiency and flow
- Elimination of the frustrations and irritations associated with dealing with traffic congestion, as well as rude and reckless drivers
- Less tiring and more enjoyable long-distance travel
- Elimination of most of the tens of thousands of fatalities and the hundreds of thousands of injuries that occur each year.

Serious obstacles and concerns must be addressed in the quest for self-driving vehicles. The problems encountered in dealing with vehicle-to-roadway and vehicle-to-vehicle interactions are complicated and formidable. During a lengthy transition period there will be a mix of vehicles with and without drivers sharing the roadways. The effects produced by the interaction of driverless cars with motorcycles, bicycles and pedestrians are unknown. Failures in the system are inevitable and most likely will result in serious delays and some spectacular collisions. Even if a driverless system is much safer and more efficient, there will be some resistance. There was widespread distrust of the "horseless carriage" even if travel by horseback, buggy and stagecoach was uncertain and extremely dangerous.[5]

These advancements would result in fewer accidents requiring investigation and reconstruction. Police officers would be relieved to focus on other public safety issues. Determining the causes, and remedies, of the accidents that do occur will require engineers and technicians knowledgeable in the design and operation of roadways, vehicles, sensors, computers and controls.

Streets and Highways

Most of our network of streets and highways fails to meet existing standards and is in a partial state of disrepair. Some is totally obsolete and presents a clear and present danger to the motoring public. One clear example is the large number of bridges that are in imminent danger of collapse.

A system of monitoring the network is necessary to identify short-term and long-term deficiencies. This would include reviewing police crash reports, monitoring user complaints and a thorough and regular system of inspection by highway agencies. It is recognized that some extra financial resources are required to properly implement a comprehensive inspection program.

Maintenance: Better maintenance of existing streets and highways is required in order to prevent further degradation. Specific maintenance problems identified by a regular monitoring system include:

- Pavement that is travel-worn, raveled, uneven, or potholed
- Pavement-shoulder drop-offs
- Overgrown grass shoulders that inhibit pavement drainage
- Clogged drainage structures
- Vegetation that blocks the view of signs or traffic signals, or violates required sight distance at intersections
- Unsafe bridges

---

[5] One state dealt with this public fear by requiring that every automobile be preceded by a responsible person carrying a red flag.

A comprehensive inspection and maintenance program is expensive but has a reasonable cost/benefit ratio.

Upgrading: Selective improvements to existing streets and highways can also produce significant improvements in safety at an acceptable cost. Relocating, changing and adding certain traffic control devices can, at relatively low cost, reduce driver-highway accident causation mechanisms. These include:

- Adding reflectorized lane lines, center lines and edge lines to improve night driving safety
- Providing advance warning of sharp curves, lane drops and merges to allow drivers sufficient time to respond
- Providing signs well in advance of intersections and interchanges informing drivers of the proper lane for each travel option in sufficient time to make the necessary lane change(s), particularly on older complex intersections and interchanges with many inconsistent features
- Adding, moving or replacing signs so that drivers are not provided information and directions faster than they can process and respond

Upgrading the roadsides has historically been a neglected area that justifies special attention since approximately 40 percent of fatalities are the result of off-road rollovers and impacts with rigid objects. Important improvements to roadside safety include:

- Adding rumble strips or rough textured shoulders
- Stabilizing soft shoulders
- Removing utility poles, trees and other rigid objects
- Replacing rigid light poles and signs with breakaway supports
- Adding energy absorbing devices to bridge supports and other immovable objects
- Installing longitudinal barriers to protect motorists from steep slopes, canals, ditches, culverts and other hazards

Renovation and Additions: The next level of improvement is to completely renovate existing streets and highways to meet existing standards for new facilities. New expressways are needed to relieve existing congestion and allow for safety improvements to existing facilities. This would be very expensive but is not an insurmountable problem as evidenced by similar undertakings in the past. In the 1950's President Eisenhower initiated the development of the Interstate Highway System that significantly improved the efficiency and safety of long distance travel. This project was very expensive but provided a big boost to the economy by improving interstate commerce and creating thousands of jobs.

Design Standards: A great deal of effort has gone into the development of highway standards including the functional design that directly impacts the driver-highway interaction. Some suggested improvements in design practices that may not meet with whole-hearted approval by some highway agencies are as follows:

- Using spiral transitions, which are basically cost free, in and out of superelevated curves with the surperelevation transition to coincide with the spiral, thus allowing for a smooth change in lateral acceleration
- Limiting the number and spacing of entrances and exits on urban expressways which means not allowing political and commercial pressure to add access that seriously degrades the safety and utility of the facility
- Permitting access and egress to freeways from the right side only
- Prohibiting the use of drainage curbs on high-speed highways since these are not redirection devices and merely contribute to vehicle damage and loss of control

Construction: Countermeasures directed toward reducing the proliferation of construction zone accidents have the potential for a significant reduction in serious and fatal crashes. Required changes to improve construction activity include:

- Requiring contractors to hire and utilize individuals trained and certified in all applicable safety standards
- Requiring highway agencies to employ inspectors that not only understand the standards but also understand the causation of construction accidents
- Levying strong penalties for non-compliance

In many areas the only effective policing of construction contractors is through civil lawsuits. Although this has had a beneficial effect, it seems to be a rather awkward and roundabout method for improving construction practices.

Injury Mitigation

Reduction in the frequency of traffic accidents can be achieved by the implementation of the countermeasures previously discussed. This would, of course, result in a corresponding elimination of a significant number of injuries and fatalities. Although collisions will still occur, related injuries and fatalities can be reduced or eliminated through improvements to highways, advances in vehicle design, and alterations to driver and passenger behavior.

Highway improvements that can mitigate injuries fall into four categories.

1. Placing longitudinal barriers to redirect vehicles away from serious roadside hazards, thus reducing the severity (smaller ΔV) of collisions.
2. Using energy absorbing (EA) devices to cushion impacts with rigid objects such as overpass supports and the termination of longitudinal barriers, thereby extending the duration of the impact with a corresponding decrease in the magnitude of the deceleration.
3. Using breakaway supports for signs and roadway lighting that dramatically reduce the change in speed during impact.
4. Removing or shielding objects, such as the ends of guardrails and the top rails of chain link fences, that can penetrate the vehicle and occupants.

Vehicle design has progressed slowly, but steadily, in no small part due to implementation of the Federal Motor Vehicle Safety standards. Some important improvements include:

1. Better door latches that reduce ejection
2. Laminated windshields, window safety glass, collapsible steering wheels, and interior padding that lessen the consequences of an occupant impacting the interior of the vehicle
3. Seat belts, air bags, head restraints, and child seats that prevent or reduce the severity of striking the vehicle interior
4. Side guard door beams that reduce intrusion into the occupant compartment during side impacts.

There are, however, still serious deficiencies in occupant protection for some common impact configurations. Door beams and side airbags provide some protection in side impacts, but intersection type collisions at normal urban and suburban speeds can result in intrusions of one to two feet that often result in serious or fatal injuries. A deadly collapse of the roof during a rollover is another example of a preventable loss of compartment integrity. The old Volkswagen Beatle was prone to a rollover, but rarely had a roof failure.

The basic approach to occupant protection should, in my opinion, have four major objectives:

1. Develop an occupant compartment that maintains its integrity during a collision. Collapse of the occupant compartment can quite effectively negate the usefulness of any restraint system.
2. Provide an easily used comprehensive restraint system of belts and airbags that prevent an occupant from striking the interior of the vehicle. This includes child seats or other systems of restraint for young children. Children and short adults should not be permitted to ride in the right front seat due to the potential for serious or fatal injuries from airbag deployment.
3. Provide interior padding to reduce injuries to restrained or unrestrained occupants who contact the vehicle interior.
4. Use design concepts that inhibit the total or partial ejection of occupants.

Another vehicle problem is created by trucks with large front or rear overhangs. Being side impacted by a truck with a large front overhang can be devastating to automobile occupants. Many trucks have long rear overhangs that can intrude into the occupant compartment of cars that rear-end them. This problem can be alleviated by providing a permanent or retractable frame under the overhang.

Drivers and passengers can have a profound effect upon their own safety by properly utilizing the available restraint systems. One argument, similar to that for motorcycle helmet laws, is that society should not have to provide for your care when you are injured – essentially underwriting your own irresponsible behavior. There are, however, other more compelling reasons for requiring drivers and passengers to "buckle up." The driver needs to be securely belted to maintain control in a relatively minor impact in order to prevent a second, possibly more severe collision. An example of this was an unbelted driver who received a non-incapacitating impact to the right side of her car, causing her to slide from behind the steering wheel. Having lost control, she crossed over a median and struck an oncoming vehicle head-on, causing not only injuries to herself, but to a totally innocent victim.

An unrestrained front or rear seat passenger can become a projectile that can incapacitate the driver or injure other restrained or unrestrained occupants. This is sufficient justification, in my opinion, for drivers and all passengers, including children, to be legally required to use the available restraint systems. Reasons for not using seat belts that I have heard include:

- It's too much trouble.
- I'm only going to the grocery.
- It will wrinkle my clothes. (Not your face?),
- I'm a careful driver (The other guy may not be.)
- What if I go into a canal?
- I am tough. I don't need a seatbelt.
- I have good reflexes. I can avoid a collision.
- I have a DEATH WISH.

Actually, I have not heard the last excuse, but I wonder sometimes if that is the real reason. One final issue is the lack of seat belts in school busses. In my opinion this not only puts children at risk, but teaches them that this is acceptable behavior.

In order for any governmental or judicial countermeasure to be effective, it must be based upon a reliable collection and interpretation of evidence, an appropriate reconstruction methodology, and a valid determination of the causal mechanism(s) justifying the countermeasure.

Appendix

Physics Review

This section is not intended as an academic or rigorous treatise. It is simply a brief *review* of some physics principles that are useful in accident reconstruction. It is assumed that the reader has had this physics background along with the mathematical capability to describe and quantify these principles.

Kinematics

Kinematics is basically the study of motion. We will use the following terminology:

t = time in seconds
d = distance in feet
d_1 = initial position
d_2 = final position
V = velocity in feet per second (ft/sec)
V_1 = initial velocity
V_2 = final velocity
a = acceleration in ft/sec^2
j = jerk in ft/sec^3.

The relationships among these variables is given by

$$V = d(\text{location})\, dt = \text{rate of change of location}$$

with respect to time at any given moment.

If V is constant

$$V = d/t$$

or

$$d = Vt\,.$$

$$a = dV/dt = \text{rate of change in velocity}$$

with respect to time at any given moment.

If a is constant

$$a = V/t \text{ or } V = at$$

$$j = da/dt = \text{rate of change of acceleration}$$

with respect to time at any moment.

If j is constant

$$j = a/t$$

or

$$a = jt\,.$$

The following forms of these relationships are also useful

$$V = \int_0^t a\, dt \text{ where } a = \text{function of time}$$

and

$$d = \int_0^t V\,dt \text{ where } V = \text{a function of time}.$$

If a is constant

$$V_2 = at + V_1$$

and

$$d_2 = \tfrac{1}{2}at^2 + V_1 t + d_1$$

If (and only if) V and a are constant

$$d = \tfrac{1}{2}at^2 \quad d = \tfrac{1}{2}Vt$$

$$a = 2d/t^2 \quad t = 2d/V$$

$$t = \sqrt{2d/a}\,.$$

Projectile Motion

As shown in Figure A.1, an object leaves point 1 with a velocity V_1 and at an angle θ_1. It lands at point 2 after traveling a horizontal distance d and with a net vertical drop h. At low speeds we can ignore air resistance with little error.

240 | HIGHWAY ACCIDENTS

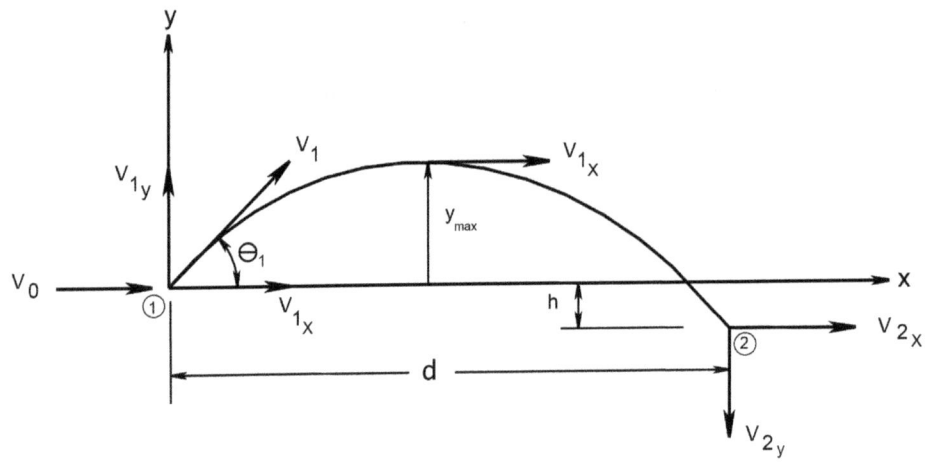

Figure A.1. Projectile Motion

Since the horizontal acceleration would be zero

$$V_x = \text{constant}$$

$$V_x = V_{1x} = V_{2x}$$

$$V_x = V_1 \cos \theta_1 .$$

The horizontal location at any time t will be

$$x = V_x t$$

$$x = V_1 \cos \theta t . \tag{A.1}$$

The vertical acceleration is a constant value g (32.2ft/sec^2) pointed downward. At any time t the vertical velocity will be

$$V_y = \int a dt$$

$$V_y = V_{1y} - at$$

$$V_y = V_1 \sin\theta_1 - gt.$$

The height above point (A.1) at any time t is

$$y = \int V_y dt$$

$$y = V_1 \sin\theta t - \tfrac{1}{2} gt^2. \tag{A.2}$$

From equation (A.1)

$$t = x/V_1 \cos\theta_1 \tag{A.3}$$

substituting (A.3) into (A.2) yields

$$y = V_1 \sin\theta_1 (x/V_1 \cos\theta_1) - \tfrac{1}{2} g(x^2/\cos^2\theta)$$

or

$$y = \tan\theta x - \frac{1}{2} \frac{g}{(V_1 \cos\theta)^2} x^2. \tag{A.4}$$

Since V_1, θ_1 and V_1, θ_1 are constants, equation (A.4) is of the form $y = bx - cx^2$ which is the parabolic projectile trajectory.

At point 2, $y = h$ and $x = d$, therefore

$$h = \tan\theta_1, d - \frac{1}{2} \frac{g}{(V_1 \cos\theta)^2} d^2.$$

If h, d, and θ_1 are known, a solution for V_1 can be obtained.

Circular Motion

The motion of an object following a circular path of radius R at a speed V is shown in Figure A.2(a). From Figure A.2(b), V_B is given by the vector sum

242 | HIGHWAY ACCIDENTS

$V_B = V_A + dV$. From Figure A.2(c), $R_A = R_B = R$. The velocity triangle and the displacement triangle shown in A.2(b) and A.2(c) are similar isosceles triangles.

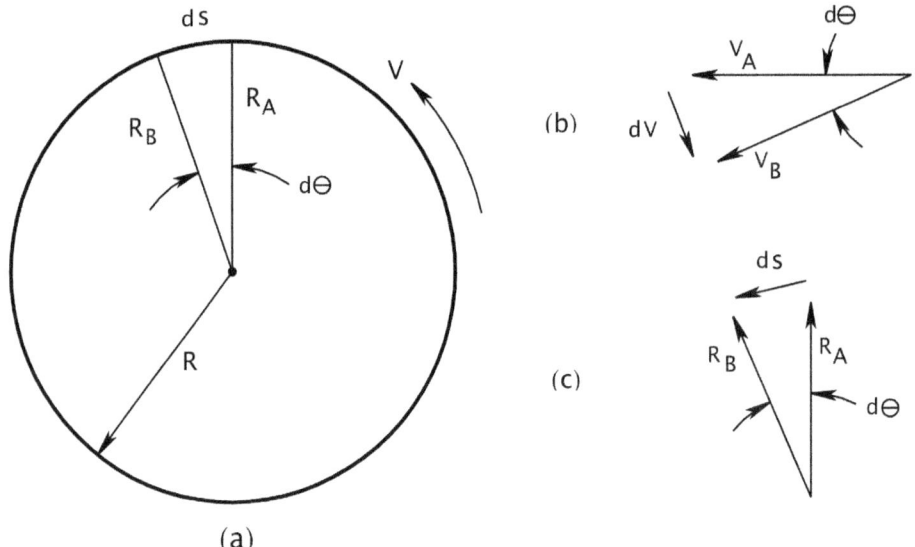

Figure A.2. Circular Motion

Therefore

$$\frac{dV}{V} = \frac{ds}{R}$$

or

$$dV = \frac{V}{R} ds.$$

Dividing by dt yields the magnitude of the acceleration

$$\frac{dV}{dt} = \frac{V}{R} \frac{ds}{dt}.$$

Since

$$a = \frac{dV}{dt} \text{ and } V = \frac{ds}{dt}$$

$$a = V^2 / R.$$

This is the centripetal acceleration that is directed toward the center of the circle and is perpendicular to the velocity vector. There may also be a tangential acceleration a_t if the speed is not constant. That is, $a_t = dV/dt$.

Dynamics

Dynamics is basically the study of motion, including the effect of forces.

Newton's Second Law is simply an observation that

$$F = ma$$

where F is the force required to give a mass m an acceleration a.

On the earth's surface the following form may be used:

$$F = \frac{W}{g} a$$

where

F = force in pounds
a = acceleration in ft/sec^2
W = weight of the object in lbs
g = acceleration of gravity (32.2 ft/sec/sec) at the earth's surface.

Work-Energy

Work is defined as the product of a force F acting through some distance d.

$$\text{Work} = Fd$$

where

F = force in lbs
d = distance in ft.

Work may also be expressed as

$$\text{Work} = (ma)d.$$

If a is *a* constant, then

$$d = \frac{1}{2}at^2$$

and

$$\text{Work} = \tfrac{1}{2}ma^2t^2$$

since

$$V = at$$

$$\text{Work} = \tfrac{1}{2}mV^2$$

where $\tfrac{1}{2}mV^2$ is also equal to the kinetic energy KE created by this work. That is

$$KE = \frac{1}{2}mV^2 \text{ or } KE = \frac{1}{2}\frac{W}{g}V^2.$$

A simple description of this relationship is that the change in kinetic energy is equal to the work done.

Power

Power is defined as the time rate at which work is done. If ΔWork is the work performed during a time interval Δt, the average power P is defined by the relationship

$$P = \Delta \text{Work} / \Delta t \,.$$

At any point in time the power P is given by

$$P = d\text{Work} / dt \,.$$

In our system of units power is given in ft.lb/sec or in horsepower (hp) where

$$1.0 \text{ hp} = 550 \text{ ft.lb/sec} = 33{,}000 \text{ ft.lb/min} \,.^6$$

Since work is equal to the change in kinetic energy, power may also be defined as

$$P = d(KE) / dt \,.$$

If P is constant

$$P = \Delta(KE) / \Delta t \,.$$

For example, a 3000 lb. car with a usable power of 100 hp accelerating from a stop for 3.0 sec would achieve a speed determined as follows

$$\Delta KE = P \Delta t$$

$$\Delta KE = 100 \ (550 \text{ ft.lb./sec})(3.0 \text{ sec})$$

$$\frac{W}{2g} V^2 = 165{,}000 \text{ ft.lb.}$$

---

[6] Our system of units was obviously designed by a committee.

$$V^2 = 2(32.2)(165{,}000)/3000$$

$$V^2 = 3542 \text{ ft}^2/\text{sec}^2$$

$$V = 59.5 \text{ ft/sec}$$

$$S = 40.6 \text{ mph.}$$

Impulse-Momentum

Momentum M is a property of a body in motion that is defined as the product of mass and velocity.

$$M = mV$$

or

$$M = \frac{W}{g}V.$$

Impulse I is defined as

$$I = \int_0^t F\,dt$$

where F is the force acting on a body over some time t.

If F is constant

$$I = Ft$$

or

$$I = mat$$

or

$$I = mV$$

where mV is the momentum created by the impulse.

Therefore, the principle expressed is that the impulse is equal to the change in momentum.

Conservation Of Energy

Energy may be in the form of potential, kinetic, heat, electrical, chemical, atomic, etc. Energy may change form but is conserved in a closed system. For example, raising an object of a weight W to a height h produces a potential energy $PE = Wh$ (i.e., the work done). Dropping the object converts the potential energy to kinetic energy equal to

$$KE = \tfrac{1}{2}mV^2$$

or

$$KE = \frac{W}{2g}V^2 .$$

Equating the two forms of energy

$$\frac{W}{2g}V^2 = Wh$$

or

$$V^2 = 2g$$

or

$$V = \sqrt{2gh}$$

where V is the terminal velocity of a dropped body.

Conservation Of Momentum

This principle is quite useful in the analysis of object-to-object impacts. It states that the total momentum of the two objects prior to impact is equal to the total momentum of the two objects after impact. This can be expressed as

$$m_1V1_1 + m_2V2_1 = m_1V1_2 + m_2V2_2 .$$

Since $m = W/g$, and g is a constant, this may also be expressed as

$$W_1 V1_1 + W_2 V2_1 = W_1 V1_2 + W_2 V2_2$$

where

W_1 = weight of object #1
W_2 = weight of object #2
$V1_1$ = velocity of object #1 prior to impact
$V2_1$ = velocity of object #2 prior to impact
$V1_2$ = velocity of object #1 after impact
$V2_1$ = velocity of object #2 after impact.

Since velocity is a vector quantity this relationship would also be true for the scalar velocity components S in a given direction. This is expressed in terms of speed S as

$$W_1 S1_1 + W_2 S2_1 = W_1 S1_2 + W_2 S2_2.$$

Friction Forces

As shown in Figure A.3, the friction force between two objects is the force resisting motion in the direction of the applied force.

F_I = applied force
F_f = resisting force
N = normal force between the objects.

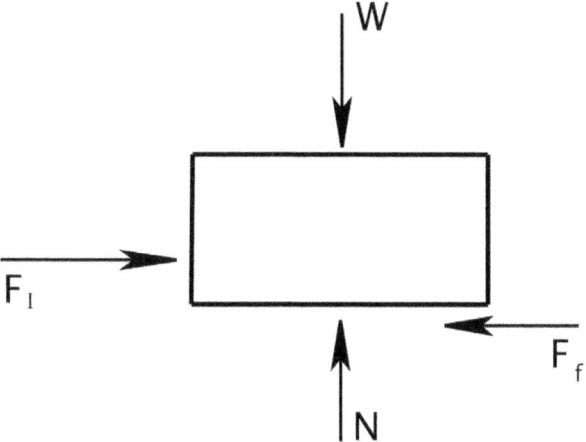

Figure A.3. Friction Force

At impending slippage the friction force is

$$F_f = f_s N$$

where

F_f = maximum friction force
f_s = static coefficient of friction.

When the objects are in relative motion (sliding) the friction force is

$$F_f = f_k N$$

where

f_k is the kinetic coefficient of friction.

In general the value of f_k is less than f_s.[7]

The kinetic coefficient of friction f_k varies significantly with speed. For example, steel sliding on steel $f_k \simeq 0.5$ at a speed of $S = 0.001$ in./sec. This reduces to $f_k \leq 0.2$ at a speed of $S = 10$ in/sec.

---

[7] This is not true for aluminum sliding on aluminum.

Moment of a Force

The moment of force about a point as shown in Figure A.4 is defined as $M_A = Fd$ $M_A = Fd$ where

M_A = the moment about point A in foot pounds (ft.lbs)
F = force in lbs
d = distance in ft.

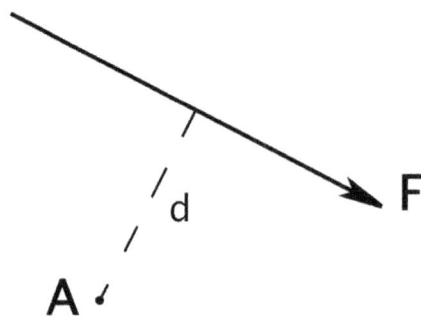

Figure A.4. Moment of a Force

Couple

A couple may be defined as two equal and opposite forces separated by a distance d as shown in Figure A.5(a). The magnitude of the couple is $M = Fd$. Since the couple has no translational effect, it is not tied to any particular point in space and could also be shown as in Figure A.5(b).

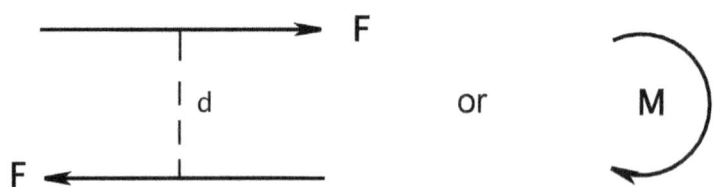

Figure A.5. Couple

Equilibrium

Static equilibrium is obtained when the vector sum of all the forces acting upon a body is equal to zero $\Sigma F = 0$, and when the sum of all moments acting upon a body are also equal to zero, $\Sigma M = 0$. Dynamic equilibrium is achieved by including the inertial force $F_I = ma$.

Rigid Bodies

The previous discussion of dynamics has essentially been limited to particle dynamics. A body with finite dimensions can experience rotation as well as translation. The kinematic relationships are as follows:

$\theta =$ rotational displacement angle in radians

$\omega =$ rotational velocity

$\omega = d\theta / dt$

$\alpha = $ = rotational acceleration

$\alpha = d\omega / dt$.

Newton's Second Law is given by

$$T = I\alpha$$

where

T = torque or moment about the axis in question
I = moment of inertia of the mass about the same axis

where I is defined as

$$I = \int r^2 dm$$

where

$r =$ perpendicular distance from the axis to the element of mass dm.

Work is defined as the torque acting through some angle θ.

That is

$$\text{Work} = \int_0^\Theta T d\theta.$$

Kinetic energy is equal to

$KE = \frac{1}{2}I\omega^2$,
power $= T\omega$,
momentum $= I\omega$ and
impulse $= \int_0^t T dt.$

A summary of various translational and rotational terms is given in Table A.1.

Table A.1. Quantities Describing Translation and Rotation

Quantity	Translation	Rotation
Displacement	$d =$ distance	θ (angle)
Velocity	$V = d(d)/dt$	$\omega = d\theta/dt$
Acceleration	$a = dV/dt$	$\alpha = d\omega/dt$
Inertia	$m =$ mass	$I =$ moment of inertia
Force or Torque	$F =$ force	$T =$ torque
Newton's Law	$F = ma$	$T = I\alpha$
Work	Fds	$Td\theta$
Kinetic Energy	$\frac{1}{2}mV^2$	$\frac{1}{2}I\omega^2$
Power	FV	$T\omega$
Momentum	mV	$I\omega$
Impulse	Fdt	Tdt

Solid Mechanics

Forces on a single object or a complex structure may cause fracture or buckling but will always produce some deformation. The resulting change in size or shape can be produced by tension, compression, bending, shear or torsion. The deformation may be temporary or permanent. If an object regains its shape when the load is released the material is defined as elastic. Materials that demonstrate no return in shape are plastic.

Impacts also exhibit elastic or plastic properties. The coefficient of restitution e is explained as follows. Drop a ball from a height h_1 onto a rigid floor. The ball will rebound to a height h_2 where $h_2 < h_1$.[8] The coefficient of restitution e is defined as

$$e = \sqrt{h_2 / h_1}\ .$$

For example, billiard balls have a coefficient of restitution approaching 1.0. Two balls of clay will show a value of ≈ 0.0. Most materials perform between these two extremes.

Materials such as mild steel exhibit both elastic and plastic properties. A steel structural component subjected to loading will behave generally in accordance with the idealized diagram shown in Figure A.6.

---

[8] If $h_2 = h_1$, you have discovered perpetual motion.

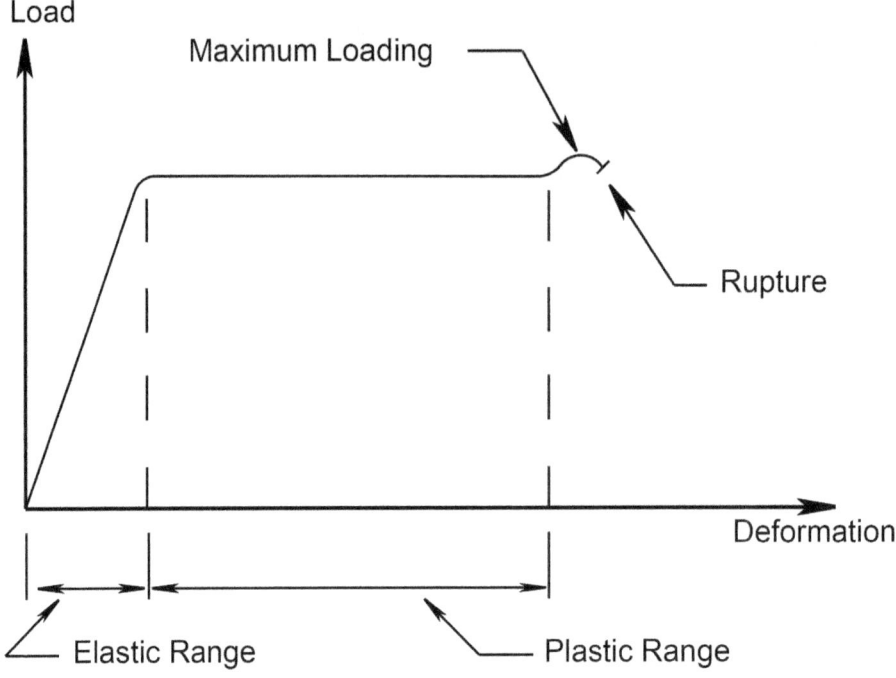

Figure A.6. Load vs. Deformation

The initial loading will produce a deformation proportional to the loading (i.e., linear). When the loading is released the component will return to its original size and shape (i.e., elastic). As the loading is increased beyond the elastic range the material will enter the plastic range. There will be a permanent deformation remaining when the loading is removed.

Spring Forces

The behavior of a linear elastic spring is shown in Figure A.7(a). This simple relationship is $F = kx$ where F is the force required to produce a deformation x and the stiffness k is the spring constant or slope of the line. The value of k can be determined experimentally by $k = F/x$.

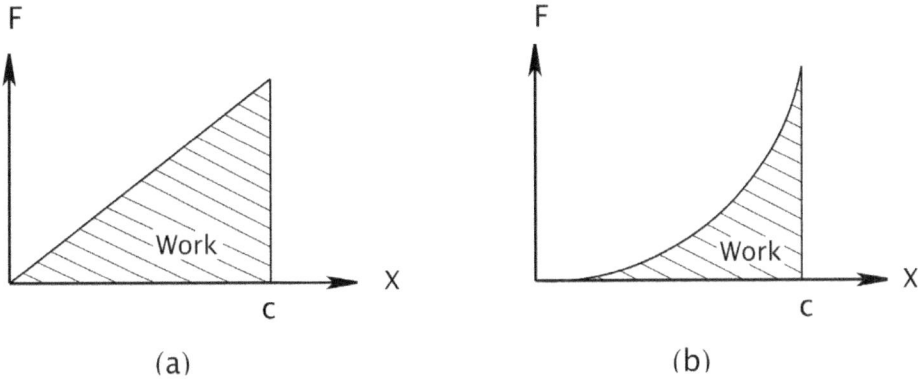

Figure A.7. Force vs. Crush

The work done in deforming the spring is equal to

$$\text{Work} = \int_0^c Fdx = \int_0^c kxdx = \frac{1}{2}kx^2 \bigg|_0^c$$

or

$$Work = \frac{1}{2}kc^2 \text{ (area under the curve)}.$$

The relationships are the same for a linear plastic spring except the deformation c is permanent.

A non-linear elastic or plastic spring will have a variable k value. The plot of force versus deformation might be similar to Figure A.7(b). The work done is equal to

$$\text{Work} = \int_0^c Fdx$$

or

$$W = \int_0^c (kx)dx$$

where the stiffness k is now a function of the deformation. That is

$$k = f(x).$$

The work done, however, is still equal to the area under the curve.